ntal Aspects of Pollution Control and Environmental Science

R.J. WAKEMAN

Department of Chemical Engineering,
University of Exeter (Great Britain)

ES
ment Contamination of the Environment

RT and R.J. STRETTON
ogical Aspects of Pollution Control

ES, H.M.A. JANSEN and J.B. OPSCHOOR
Approaches to Environmental Problems

ROD
in Horticulture

N
in the Soil Environment

s in this series (in preparation):

EY and R.E. REDMANN
change in Ecosystems

RT
Desulfurization

IQI and F.L. WORLEY, Jr.
on Measurements and Monitoring

ON
of Solutes into Groundwater

Fundamental Aspects of Pollution Control a

PESTICIDES IN THE SOIL E

Fundame

Edited by

1
D. PURV
Trace-Ele

2
R.K. DA
Microbiol

3
D.E. JAM
Economi

4
D.P. ORM
Pollution

5
S.U. KHA
Pesticides

Other title

R.E. RIPI
Energy Ex

W.L. SHO
Flue Gas I

A.A. SIDI
Air Polluti

D.R. WILS
Infiltratio

Fundamental Aspects of Pollution Control and Environmental Science 5

PESTICIDES IN THE SOIL ENVIRONMENT

SHAHAMAT U. KHAN
Chemistry and Biology Research Institute
Research Branch, Agriculture Canada
Ottawa, Ont., Canada

ELSEVIER SCIENTIFIC PUBLISHING COMPANY
Amsterdam — Oxford — New York 1980

ELSEVIER SCIENTIFIC PUBLISHING COMPANY
335 Jan van Galenstraat
P.O. Box 211, 1000 AE Amsterdam, The Netherlands

Distributors for the United States and Canada:

ELSEVIER/NORTH-HOLLAND INC.
52, Vanderbilt Avenue
New York, N.Y. 10017

Library of Congress Cataloging in Publication Data

Kahn, Shahamat U
Pesticides in the soil environment.

(Fundamental aspects of pollution control and environmental science ; 5)
Includes bibliographical references and indexes.
1. Pesticides--Environmental aspects. 2. Soil pollution. I. Title. II. Series.
TD879.P37K33 631.4'1 80-11238
ISBN 0-444-41873-3

ISBN 0-444-41873-3 (Vol. 5)
ISBN 0-444-41611-0 (Series)

Printed in The Netherlands

PREFACE

Chemicals for crop protection and pest control - known collectively as pesticides - are being increasingly used to ensure the production of adequate supplies of food and fiber. Some of these pesticides find their way into soils as a result of direct application or through indirect means. With the discovery that chlorinated hydrocarbon insecticides persist for years in soil, all pesticides are now being viewed with suspicion and concern by people interested in protecting our agricultural land from widespread pollution.

The extent and seriousness of the contamination of soils by pesticides still remains to be determined. Some environmentalists take the view that use of pesticides on agricultural soils should be reduced or banned because of the risk of uptake of these chemicals by crops and their subsequent incorporation into the food chain. On the other hand, agriculturalists and others argue that continued use of large quantities of pesticides is essential to the achievement of maximum yields. A reasonable alternative to these extreme views would be to first gain a better understanding of the behavior of pesticides in soils from the standpoint of the processes affecting these chemicals, and the implication of these processes on persistence, bioactivity and plant uptake. With this knowledge, the environmental impact of using a pesticide in agriculture could be assessed more accurately. This book, *Pesticides in the Soil Environment*, is an attempt to provide this kind of information by bringing together the available data on many aspects of the behavior and fate of pesticides in soils. It is hoped that it will serve as a text book for advanced courses, a reference volume for research workers and a source of detailed information for those who seek knowledge on the topic.

I will make no effort to acknowledge individually the many people who assisted me in proof reading, in the preparation of illustrations and the compilation of the indexes. To them I am grateful. I do wish, however, to express my appreciation to Mrs. Anneth Martin for her painstaking efforts in the final typing of the manuscript. My sincere gratitude is also expressed to the Chemistry and Biology Research Institute, Research Branch, Agriculture Canada, for providing opportunity and facilities to produce this book.

Finally, I must convey my deepest affection and appreciation to my wife Nighat and to my children, Saira and Zia, for their keen sense of understanding during the preparation of this book.

Ottawa, Ontario
December, 1979

Shahamat U. Khan

THE AUTHOR

SHAHAMAT U. KHAN is a Senior Research Scientist at the Chemistry and Biology Research Institute, Research Branch, Agriculture Canada, Ottawa. His research is concerned with the fate of pesticides in the environment.

He obtained a B.Sc. in Pure Science from Agra University, India, an M.Sc. in Chemistry from Aligarh University, India, and an M.Sc. and a Ph.D. in Soil Chemistry, both from the University of Alberta, Edmonton, Canada.

Dr. Khan belongs to numerous scientific societies and is a Fellow of the Chemical Institute of Canada and a Fellow of the Royal Institute of Chemistry (London). He is the Editor of the *Journal of Environmental Science and Health, Part B*.

He is the author or coauthor of more than 80 scientific research publications and has coauthored a previous book, *Humic Substances in the Environment* (1972) and coedited another book *Soil Organic Matter* (1978). In addition he has written a number of chapters in edited books and several review articles.

CONTENTS

Chapter 1

INTRODUCTION

Man has practiced some form of pest control since the beginning of agricultural times. The principles of seed treatment, fumigation and the use of certain preparations to kill unwanted pests were known to the ancient agriculturalists. Only in the last thirty years, however, has the use of chemical agents produced substantial benefits for mankind. Pesticides have controlled weeds, pests infesting economically important crops, vectors of human and animal diseases and have protected structures from damage. As the world's population increases so does the need for food and fiber production. Crop protection and pest control should therefore be continued and intensified.

Chemicals classified as pesticides have been used to some extent since ancient times. Arsenic was used by the Chinese in A.D. 900 to control garden insects. During the 17th century arsenic and tobacco were used as insecticides in the Western world. Beginning about 1870 the number of compounds available for use as pesticides increased gradually and equipment for applying these chemicals began to be developed. A recognizable acceleration in the rate of the introduction of pesticides began in 1924 with a still further increase in 1946. Some important insecticides were discovered during World War II, but these discoveries had far less to do with the war *per se* than is commonly assumed. Over the past three decades, increases in crop yields have largely been due to the production and use of enormous quantities of pesticides each year. The development of chemicals for crop protection can be attributed almost entirely to the pesticide industry. The phenomenal growth rate of the world pesticide industry over the past three decades is illustrated in Fig. 1.1. The value of pesticides produced in the world in 1974 is shown in Fig. 1.2 (Green et al., 1977). In 1971 \$3.4 billion worth (retail) of chemical pesticides was applied on a world wide basis for agricultural (including forestry), industrial, and household use

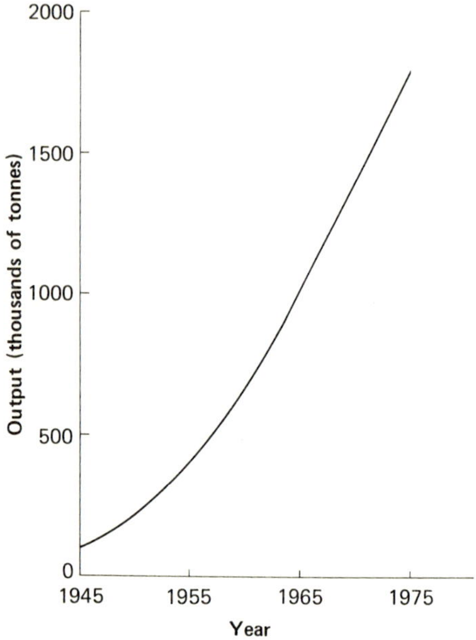

Fig. 1.1. Growth of world pesticide industry.

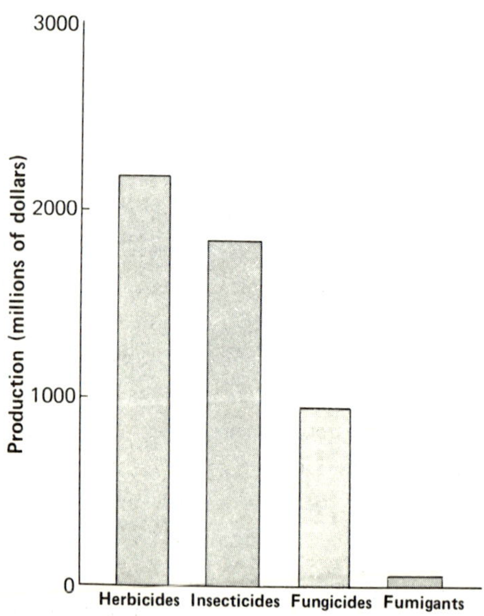

Fig. 1.2. Value of pesticides produced in the world in 1974.

(Anon., 1973). About half was used in the United States, where pesticide consumption has upsurged notably in the past 30 years. It is certain that the demand for pesticides will increase as the human population and its food and fiber requirements continue to grow. Table 1.1 shows the projected world demand and market forecast for pesticides based on price levels of the year 1975 (Green et al., 1977).

TABLE 1.1

Forecasts of world demand for pesticides

Chemical	Millions of dollars		
	1975	1980	1990
Herbicides	2300	3450	7700
Insecticides	1910	2390	3700
Fungicides	1035	1345	1880
Total	5245	7185	13280

In recent years the use of pesticides has grown impressively despite rising prices. For instance, in the United States the average value of all chemicals classified as pesticides increased at an average annual rate of 15.9% for the five year period 1972 to 1977, while sales of pesticides rose at an average annual rate of 26.3% for the same period (U.S. Dept. Agric., 1977). It is apparent that, in spite of increases in price, the use of pesticides can be expected to grow as an economic necessity. Pimental (1973) estimated that a $10 billion average loss in the United States in 1960 would have increased to $12 billion had pesticides not been applied. The cost of such pesticides, in 1966 for example, was $0.56 billion. Including application, the total cost was about $0.75 billion, representing nearly $3 saved for every $1 spent.

Despite the widespread use of pesticides, the U.S. Department of Agriculture estimated that in 1971, the agriculture industry in the United States alone, absorbed a loss of $10 billion annually owing to insects, weeds, plant diseases and nematodes. On the world level the losses to pest, plant diseases and weeds were estimated to exceed $70 billion (Marmet, 1977). Crop losses in less developed countries are judged to be greater than those in the industrialized nations. Almost one half of the potential food production of the less developed countries in the tropics is lost due to the ravages of insects, plant disease organisms, weeds, rodents, birds, nematodes and others (Table 1.2). It has been estimated that cessation in the use of all pesticides in the United States would reduce total production of all crops and livestock by 40% and increase the price of farm products to the consumer by 50 to 70%.

TABLE 1.2

Losses of potential crop production by region (Glass, 1976)

	% losses due to			
Regions	Insect pests	Diseases	Weeds	Value of lost production $ millions
North and Central America	9.4	11.3	8.0	9837
South America	10.0	15.2	7.8	4561
Europe	5.1	13.1	6.8	11927
Africa	13.0	12.9	15.7	7735
Asia	20.7	11.3	11.3	27290
Oceania	7.0	12.6	8.3	476
USSR and People's Republic of China	10.5	9.1	10.1	8521

According to the census of Agriculture in the United States for 1974, the average cost of controlling pathogens was $20.03 per treated acre; nematodes, $16.50; insects in crops other than hay, $10.87; weeds in crops, $7.08; weeds in pasture, $3.17; insects in hay, $5.82; and for plant defoliation, $6.65 (U.S. Dept. Agric., 1977). Many farmers have been willing to spend money on pesticides because the investment has been profitable for them. It has been estimated that each dollar spent on pesticides in the United States produces an average of about $4 additional income for the farmer. It is, however, not possible to predict the value of the use of pesticides to individual farmers because of wide variations in types of crops, geographical locations, climatic conditions and the skill with which the chemicals are used. An optimistic view is that the increased use of pesticides will prove profitable for farmers and will contribute substantially to increases in yields per acre and per man hour for all major crops.

It would be incorrect to imply that the use of pesticides in crop protection is free from problems. Pesticide residues may constitute a significant source of contamination of air, water, soil and food, which could become a threat to the continued existence of many plant and animal communities of the ecosystem. The continual addition of large amounts of persistent pesticides to the environment has caused great anxiety among many ecologists. A variety of undesirable environmental effects of pesticides has been reported from many countries. The effects include excessive mortality and reduced reproductive potential in organisms such as birds and fish; changes in the abundance of species and the diversity of ecosystems; a reduction in the productive potential of natural resources and the development of pesticide resistance in target and nontarget species (Koeman, 1978).

Regardless of the method of application, large amounts of pesticides ultimately reach the soil. As a result, world soils are accumulating ever increasing amounts of residues of a wide variety of pesticides which then move into the bodies of invertebrates, pass into air or water, are absorbed by plants, or are broken down into other products. The presence of pesticides in the soil must therefore continue to be of major interest to environmental scientists.

The purpose of this book is to highlight many aspects of pesticides in the soil environment. The conventional classification of the pesticides into herbicides, insecticides, and fungicides has been followed for the most part in this book. Hopefully this will provide an adequate representation of the different classes of chemicals and so illustrate various aspects of the fate of pesticides in soil. A complete catalogue of the structures and properties of all chemicals in current use as pesticides was not possible in the space available without drastic restriction of other desirable material.

The behavior and fate of pesticides in the soil is discussed in terms of physicochemical and microbiological processes. In order to understand the precise nature of the physicochemical processes involved, numerous interactions between pesticides and soil constituents are discussed in chapter 3. This same chapter also includes a discussion of the movement, volatilization, photodecomposition, chemical conversion and degradation of pesticides in soil. These physicochemical processes play an important role in the dissipation of pesticides in soil and help in the prediction of the probable effectiveness of the chemicals in pest control.

The biochemical reactions associated with the microbial metabolisms of various classes of pesticides are discussed in chapter 4. The processes by which pesticides undergo degradation are examined and microbial involvement is identified. Ideally, the pesticides should remain active long enough to accomplish the intended task, then decompose to innocuous products before another application becomes necessary. However, persistence of the pesticides beyond the critical period for control leads to residue problems. Chapter 5 brings together much of the available data on the occurrence and persistence of pesticide residues in soils. The uptake of residues by plants and soil animals is also discussed in this chapter. It is important that crops used for human and animal food should not contain any residues of pesticides. In many countries legal limits or tolerances have been established for the amounts of pesticide residues that are permissible in plant tissues to be used for food. A problem more complex than that of the toxicity of pesticides to soil animals is the accumulation of residues in their body tissues. This raises concerns as to whether animals and birds feeding upon these invertebrates will concentrate these residues even further. Chapter 5 also includes a discussion on the nature, significance and the source of bound residues in soil. Specific attention has been given to the critical question of qualitative and quantitative determination of bound residues and their biological availability. It is conceivable that a change in cultural practices may liberate bound residues and reintroduce them into the soil solution, which may subsequently result in their being taken up and translocated into the economic portions of plants.

The last chapter presents a brief account of the complex problem concerned with minimizing pesticide residues in soil. Pest control methods that do not require the use of pesticides, such as biological control, as well as the possibility of using short residual pesticides with narrow spectra of toxicity are briefly discussed. The chapter concludes with a short discussion of the continuing need of chemicals for crop protection and pest control in the foreseeable future.

The author has chosen not to include information on the numerical changes induced by pesticides on soil microorganisms. Furthermore, no attempt has been made to discuss the effects of pesticides on the chemical and physical properties of soil. This omission was necessary in order to adequately cover the pesticides-soil aspects within the available space. In addition, the topics in this book have been selected with the primary aim of presenting as balanced a picture as possible of the present status of the fate and behavior of pesticides in the soil environment.

REFERENCES

Anonymous, 1973. Farm Chemicals and Croplife, 136:26-30.
Glass, E.H., 1976. National Technical Information Service Report PB-257 361, Ithaca, N.Y., 70 pp.
Green, M.B., Hartley, G.S. and West, T.F., 1977. Chemicals for Crop Protection and Pest Control, Pergamon Press, New York, N.Y., 291 pp.
Koeman, J.H., 1978. In: Advances in Pesticide Science, Part I., H. Geissbühler (Editor), Pergamon Press, New York, N.Y., pp. 25-38.
Marmet, J.P., 1977. Pestic. Sci., 8:380-388.
Pimental, D., 1973. J.N.Y. Entomol. Soc., 81:13-33.
United States Department of Agriculture, 1977. The Pestic. Rev., Washington, D.C., 44 pp.

Chapter 2

CLASSIFICATION OF PESTICIDES

A pesticide can be defined as any substance or mixture of substances intended for preventing, destroying or repelling any insect, nematode, fungus, insect, weed or any other form of terrestrial or aquatic plant or animal or microbiological life, and for use as a plant regulator, defoliant or desiccant. The chemicals represent many different classes of compounds and are usually grouped according to the purpose for which they are used. In agriculture, herbicides, insecticides and fungicides are used for controlling weeds, insects, and plant pathogens, respectively. It is not the purpose of this chapter to describe the many details of the existing pesticidal compounds such as their use, characteristics, and commercial value. Rather, the intention is to describe briefly only those pesticides that may eventually enter the soil environment by their application directly to soil or by aerial or foliar spray. Most of the information given in this chapter has been reviewed elsewhere (Crafts, 1961; Metcalf, 1971; Brooks, 1974; Eto, 1974; Khur and Dorough, 1976).

Direct application of pesticides may result in an accumulation of their residues in soil. A large proportion of foliar sprays that do not reach their target may also contribute greatly to soil residues. Pesticides may also reach the soil when leaves that have been sprayed fall to the ground or crops that contain small amounts of pesticides are ploughed in or when bodies of animals with residues in their tissues are buried. Another source of pesticides in soil is the residues of these chemicals in the atmosphere, either in dust or rain water, which can be washed out by precipitation and fall onto the soil.

2.1. HERBICIDES

Herbicides available to the farmer contain compounds of widely differing physical, chemical and biological properties. Some herbicides are applied directly to the soil to achieve weed control, whereas others are used primarily as foliar applied treatments. In the latter case, varying amounts of the chemical reach the soil. A variety of methods have been used for herbicide application to the soil. The most widely used technique is that of soil incorporation, which minimizes volatilization. Other techniques include subsurface and sequential applications and application before planting. Herbicides can be applied in a formulated form. For example, granular formulations can be prepared to regulate volatilization and leaching, while the choice of solvent, surfactant, and water proofing agents can control the release of the chemical. Because of synergistic effects, application of a mixture of herbicides may result in the use of lesser amounts of chemicals than would be required if the components were applied separately. This may reduce side effects from use of the individual chemical at a higher rate. The herbicides are classified by grouping the compounds chemically.

2.1.1. Arsenicals

Sodium arsenite (1) has been used as a weed killer on railroad right-of-ways in the United States, and in sugar cane and rubber plantations in tropical countries. Cacodylic acid (2) and its sodium salt (3) have been found useful as general contact herbicides to control weeds. Another organic arsenical compound,

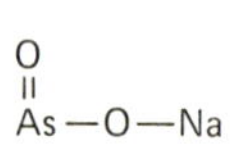

1

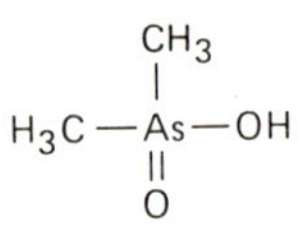

2

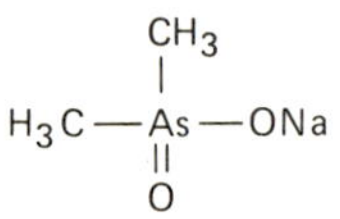

3

namely disodium methanearsonic acid (4) is still used on a large scale.

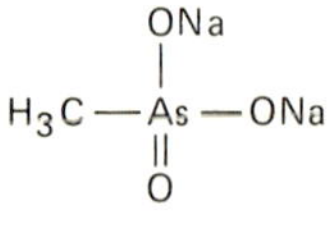

4

2.1.2. Organophosphates

A number of organophosphorus compounds show herbicidal activity. Commercially used compounds include DMPA (5) amiprophos (6) and metacrephos (7).

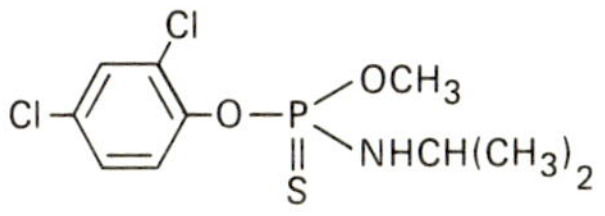

5

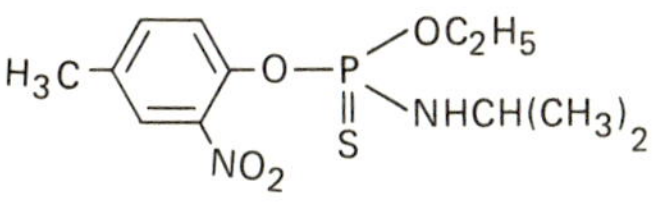

6

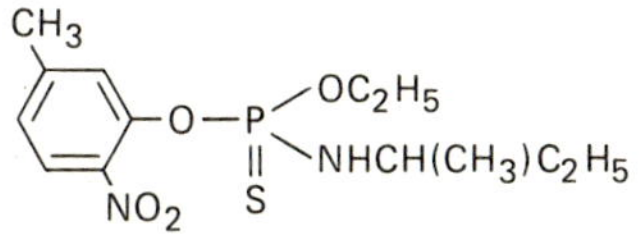

7

Glyphosate (8) is a very broad spectrum and by far the most important organophosphorus herbicide. It is a contact herbicide active only for foliar application.

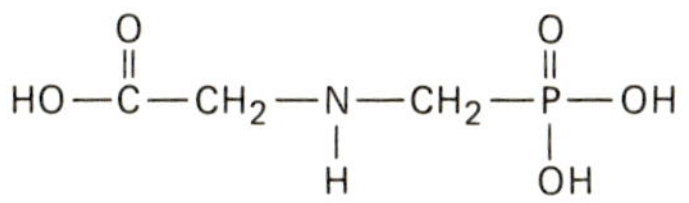

8

2.1.3. Phenoxys

The chlorinated phenoxy acids have been the key herbicides for the very rapid expansion of chemical weed control in the last 30 years. They are selective to broad leaved weeds in cereals and grasses. They are used as herbicides in the form of the parent acids, as salts and as esters. The most widely and commercially used compound of this family is 2,4-D (9). Two other important compounds are an ester of 2,4-D, MCPA (10) and the closely related compound 2,4,5-T (11).

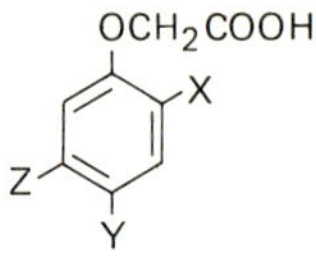

X, Y=Cl; Z=H 9

X=CH_3; Y=Cl; Z=H 10

X, Y, Z=Cl 11

2.1.4. Benzoics

These compounds are especially useful for the control of deep rooted perennial weeds. Those developed into commercial products include 2,3,6-TBA (12), dicamba (13), tricamba (14), chloramben (15).

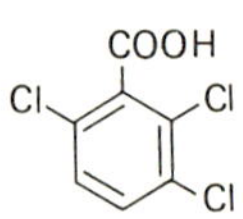

12

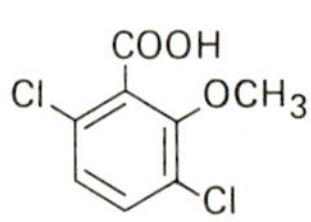

13

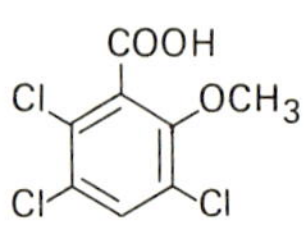

14

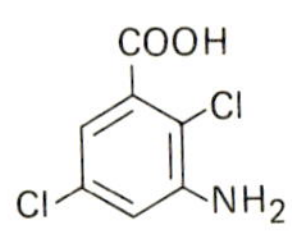

15

2.1.5. Pyridine Acids

Picloram (16) is a systemic herbicide and controls broad leaved weeds. This is the only prominent member of the family of pyridine derivatives that has been studied extensively and developed commercially as a herbicide.

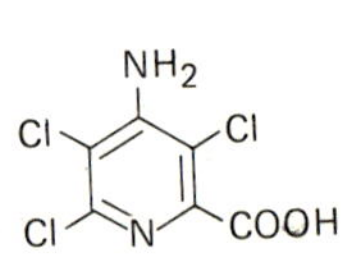

16

2.1.6. Chlorinated Aliphatic Acids

The two commonly used herbicides TCA (17) and dalpon (18) are effective against grasses. Although they are often referred to as chlorinated acids, many are used almost exclusively as the sodium salts.

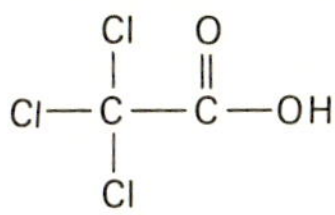

17

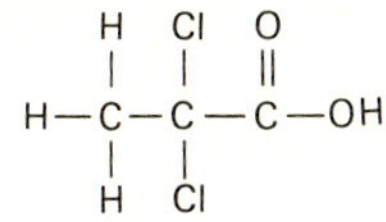

18

2.1.7. Amides

These compounds are almost exclusively used as selective herbicides in a variety of crops. They range from such structures as *N*-substituted α-halo acetamides and α,α-diacyl acetamides through substituted aromatic anilides of aliphatic acids and cyclopropyl carboxylic acids to *N*-naphthalamic acids. The most commercially successful compound has been propanil (19). Other herbicides of great commercial utility include propachlor (20) and alachlor (21).

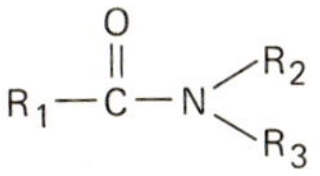

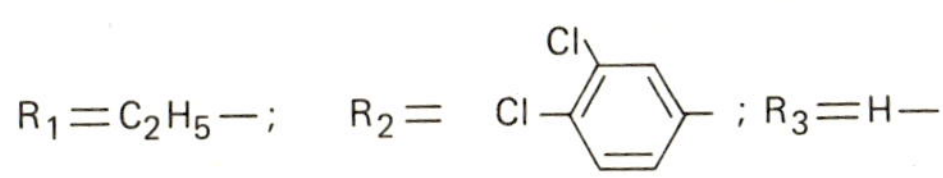

19

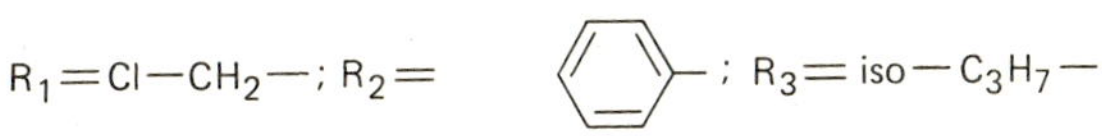

20

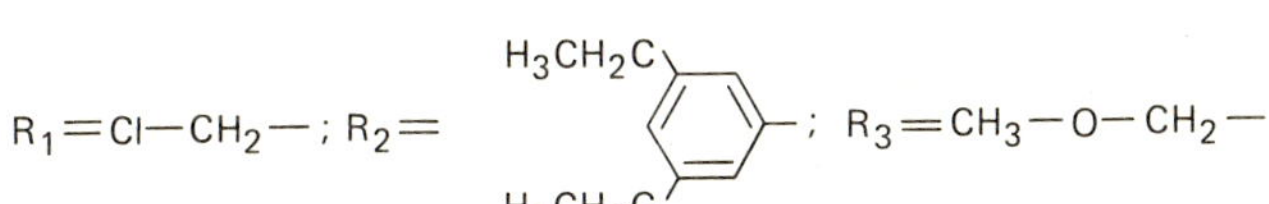

21

2.1.8. Carbamates and Thiocarbamates

The carbamate herbicides are becoming increasingly important because of their low mammalian toxicity, relatively short residual life in soil, and degradation by nontarget organisms. These herbicides derive their basic structure from carbamic acid (22). A wide range of carbamate herbicides are now available to give broad spectrum weed control. The most commercially used compounds are chlorpropham (23), Swep (24), propham (25), and barban (26).

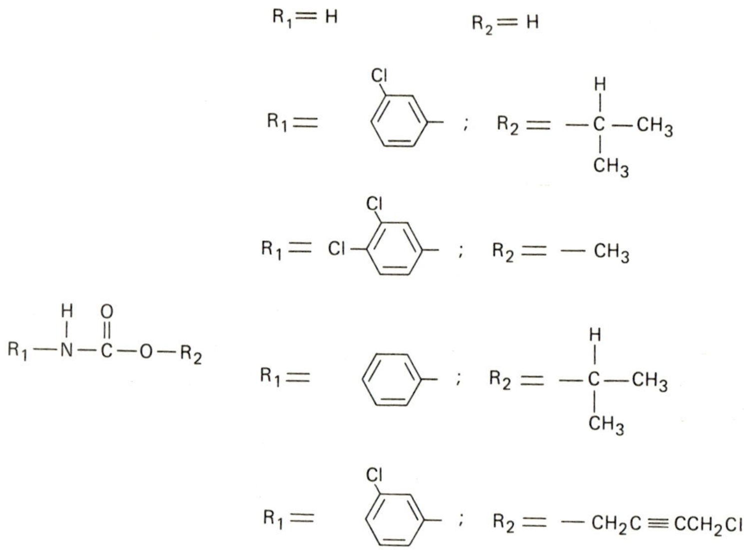

22

23

24

25

26

A number of *N*-alkylthiocarbamates are of interest among pest control chemicals. Substitution of one sulfur atom for an oxygen in carbamic acid (22) gives thiocarbamic acid (27), and two sulfur substitution gives dithiocarbamic acid (28). Derivatives

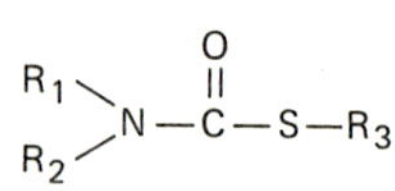

$R_1 = R_2 = R_3 = H$ 27

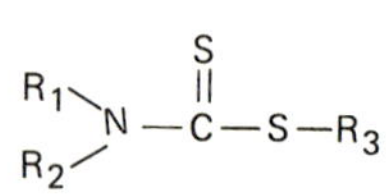

$R_1 = R_2 = R_3 = H$ 28

of thiocarbamic acid (27) include diallate (29), triallate (30), EPTC (31) and vernolate (32).

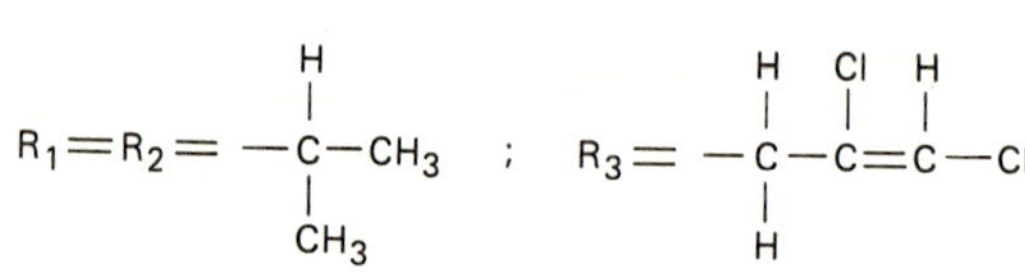

29

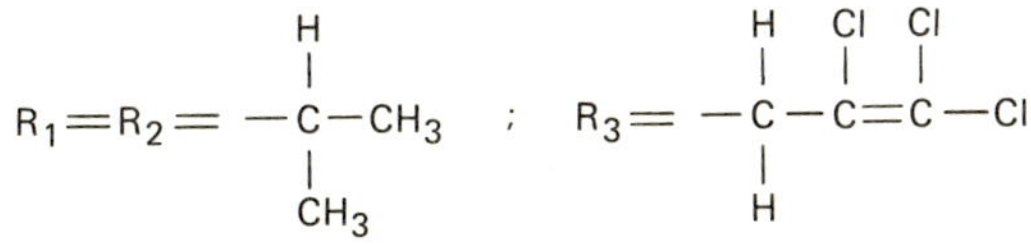

30

$R_1 = R_2 = -C_3H_7$; $R_3 = -C_2H_5$ 31

$R_1 = R_2 = R_3 = -C_3H_7$ 32

The two dithiocarbamate compounds, metham (33) and CDEC (34), which are used as herbicides, are the derivatives of dithiocarbamic acid (28).

$R_1 = -CH_3$; $R_2 = -H$; $R_3 = -Na$ 33

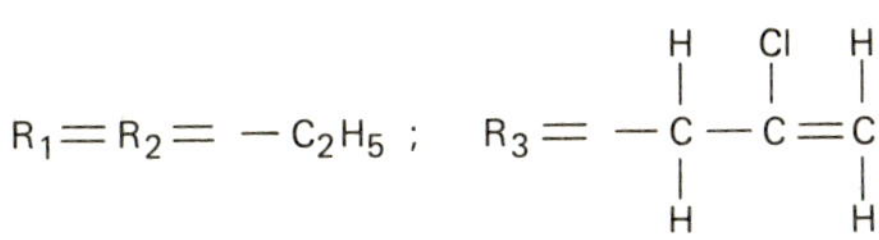

34

2.1.9. Dinitroanilines

These herbicides are generally used for selective weed control as a preplanting soil incorporation treatment prior to weed germination. The 2,6-dinitroanilines possess a marked general herbicidal activity. Substitution at the 3 and/or 4 position of the ring or on the amino group modifies the degree of herbicidal activity. However, it does not essentially change the type of herbicidal activity provided that the 2,6-dinitroaniline structure is retained. The commonly used herbicides in this group include trifluralin (35), benefin (36), nitralin (37) and dinitramine (38).

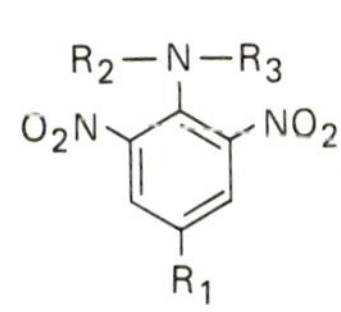

$R_1 = -CF_3 ; R_2 = R_3 = -C_3H_7$ 35

$R_1 = -CF_3 ; R_2 = -C_2H_5 ; R_3 = -C_4H_9$ 36

$R_1 = -SO_2CH_3 ; R_2 = R_3 = -C_3H_7$ 37

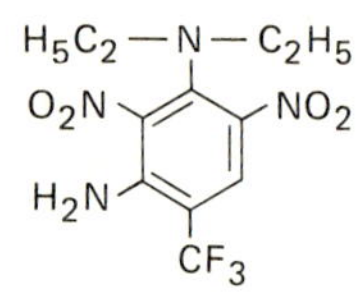

38

2.1.10. Nitriles

These compounds have proved useful in controlling annual weeds and broadleaf weeds that sometimes do not respond to 2,4-D (9). These herbicides are of rather recent development and are exemplified by dichlobenil (39), ioxynil (40) and bromoxynil (41).

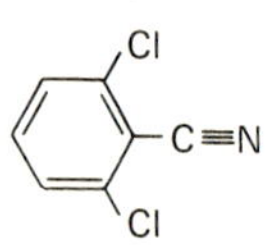

39

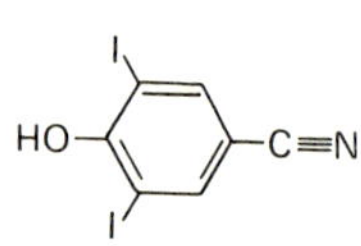

40

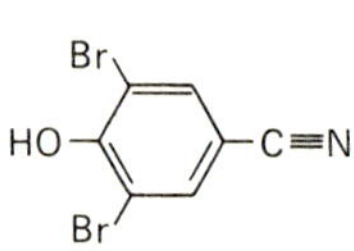

41

2.1.11. Phenols

The broad spectrum of activity of some substituted phenolic herbicides has fostered their use against broadleaf annual weeds in many crops. The most commercially important herbicides in this group are dinoseb (42), DNOC (43), dinosam (44) and PCP (45).

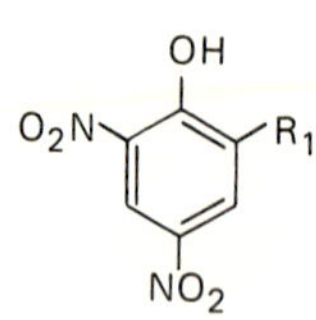

$R_1 = -CHCH_3C_2H_5$ 42

$R_1 = -CH_3$ 43

$R_1 = -CHCH_3C_3H_7$ 44

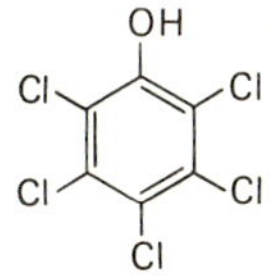

45

2.1.12. Bipyridyliums

The bipyridylium compounds are usually used as general contact weed control agents and are nonselective, quick acting herbicides and desiccants. Diquat (46) and paraquat (47) are the most important heterocyclic organic compounds used as herbicides. They are available commercially as dibromide or dichloride salts.

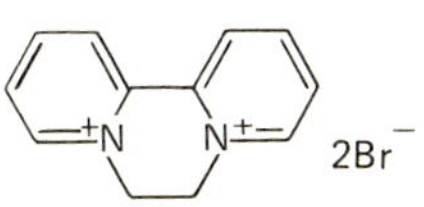

46

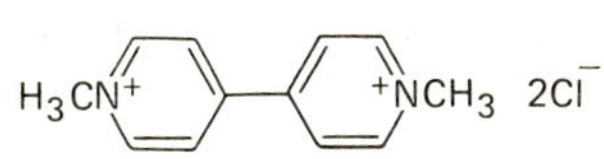

47

2.1.13. Uracils

These herbicides are related to the pyrimidine bases. They are used for general weed control in non crop land and are particularly effective against perennial grasses. Three of the substituted uracil herbicides most commonly used are isocil (48), bromacil (49), and terbacil (50).

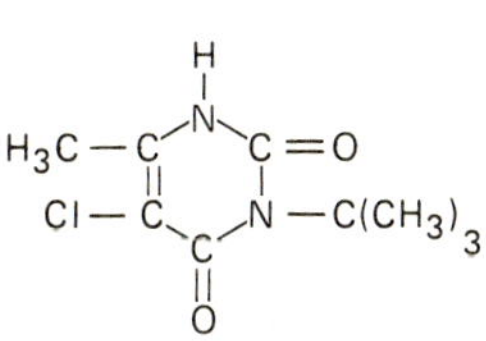

48

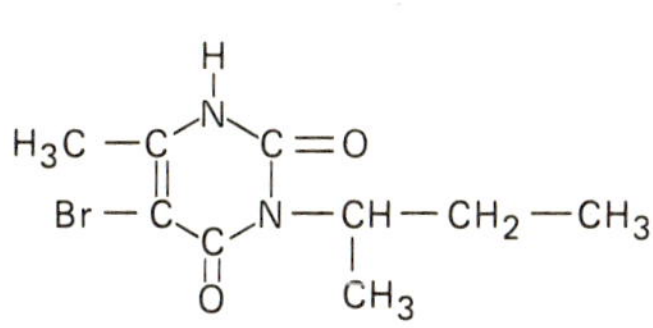

49

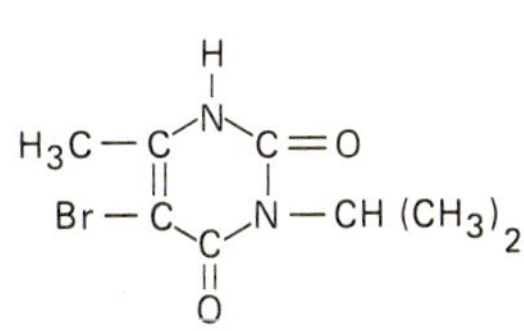

50

2.1.14. Triazoles

The commercially used herbicide amitrole (51) is currently not registered for use in any crop in the United States. However, it is being used for weed control in noncropped areas.

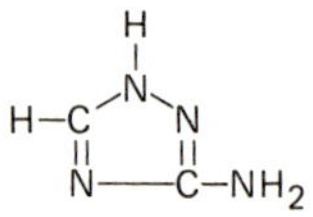

51

2.1.15. *s*-Triazines

In recent years the *s*-triazines have become one of the most important and widely used group of herbicides. They are used as selective as well as nonselective herbicides. Atrazine (52), is the herbicide which found a major use in agriculture. Other *s*-triazines commercially used in agriculture include simazine (53), prometryn (54) and ametryn (55). Recently, several other *s*-triazines, such as prometone (56), propazine (57), and simetone (58) were introduced to the market.

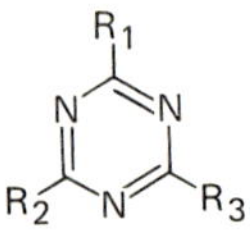

$R_1 = -Cl$; $R_2 = -NH \cdot iso-C_3H_7$; $R_3 = -NHC_2H_5$ 52

$R_1 = -Cl$; $R_2 = R_3 = -NHC_2H_5$ 53

$R_1 = -SCH_3$; $R_2 = R_3 = -NH \cdot iso-C_3H_7$ 54

$R_1 = -SCH_3$; $R_2 = -NH \cdot iso-C_3H_7$; $R_3 = -NHC_2H_5$ 55

$R_1 = -OCH_3$; $R_2 = R_3 = -NH \cdot iso-C_3H_7$ 56

$R_1 = -Cl$; $R_2 = R_3 = -NH \cdot iso-C_3H_7$ 57

$R_1 = -OCH_3$; $R_2 = R_3 = -NHC_2H_5$ 58

2.1.16. Ureas

The most important compounds developed in this group for commercial application include linuron (59), diuron (60), monuron (61), fenuron (62), and neburon (63). Most of these herbicides are relatively nonselective and are directly applied to the soil; however, some are active through the foliage. In addition to the compounds shown below, several other urea herbicides are also available commercially. Diuron (60) is by far the most commercially useful urea herbicide.

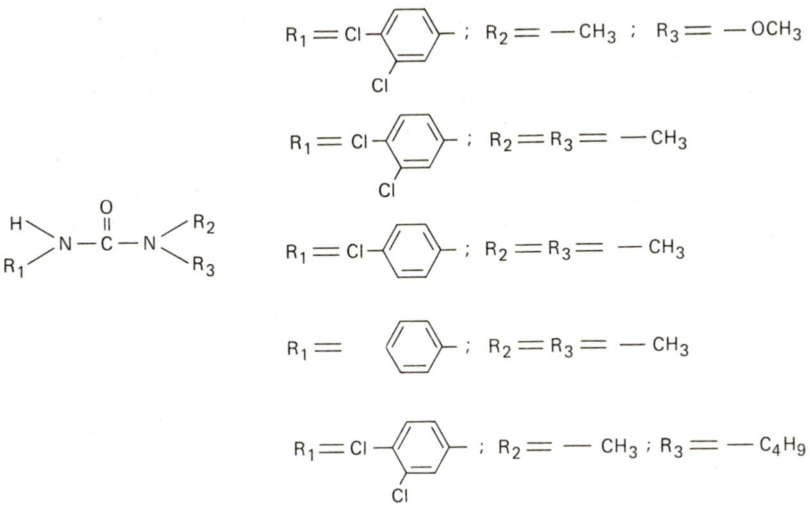

59

60

61

62

63

2.2. INSECTICIDES

The three important classes of insecticides are the organophosphorus compounds, carbamates and chlorinated hydrocarbons. They are usually applied directly to the soil to kill soil borne pests. When applied as aerial sprays or dust to foliage, a large amount of them also ultimately reach the soil. Insecticides have been broadcast over the surface of soil and then thoroughly incorporated into the soil with a plough. Unfortunately, such treatments may result in using much more insecticide than is really necessary to control a particular pest in the soil. In some cases, such as seed dressing, it may be preferable to use localized treatment to place the insecticide exactly where it is required. Other application techniques include the gradual release of the chemicals into the soil from microcapsules or from the surface of innert granules.

2.2.1. Organophosphorus Compounds

The organophosphorus insecticides are hydrocarbon compounds which contain one or more phosphorus atoms and are relatively short lived in biological systems. They are soluble in water and readily hydrolyzed. Many organophosphorus pesticides dissipate from soil within a few weeks after application. Because of their low persistence and high effectiveness, these compounds are now used widely as systemic insecticides for plants, animals, and for seed and soil treatments. The organophosphorus insecticides are used as stomach and contact poisons, as fumigants, and as systemic insecticides for nearly every type of insect control. In this section a brief description will be given for only those commercially important organophosphorus insecticides that are used in the soil.

2.2.1.1. Phosphates

Most of the commercial products are vinyl ester derivatives of phosphates such as dichlorvos (64) chlorfenvinphos (65), mevinphos (66), crotoxyphos (67) and dicrotophos (68).

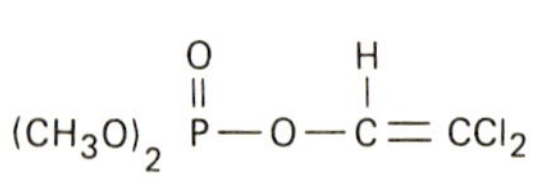

64

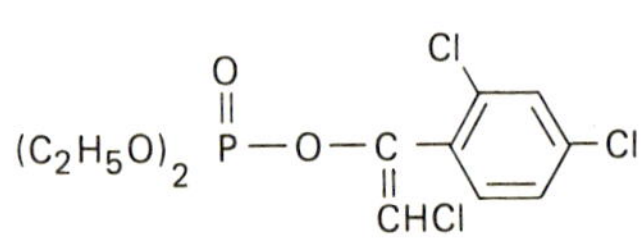

65

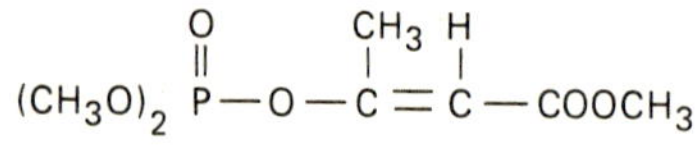

66

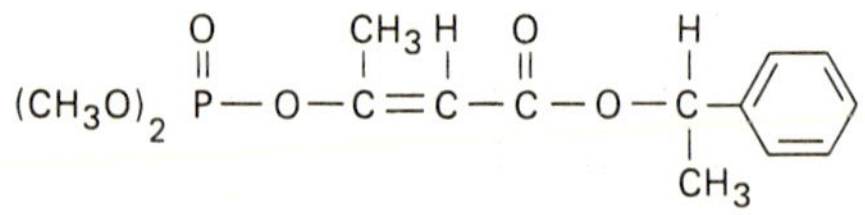

67

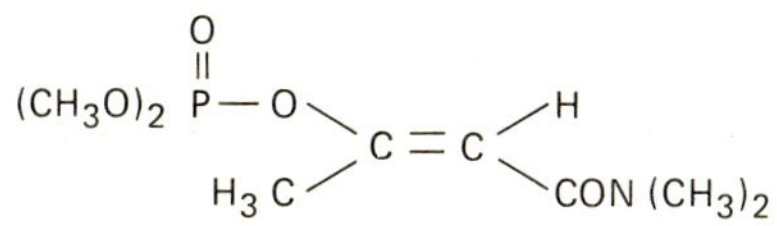

68

2.2.1.2. Phosphorothioates

In general, these insecticides have a greater hydrolylic stability under aqueous conditions and are usually more active as insecticides than the corresponding phosphate analogues. The most widely used compounds are parathion (69), parathion methyl (70), diazinon (71), dursban (72), fenitrothion (73), fenthion (74) and demeton-*O* (75).

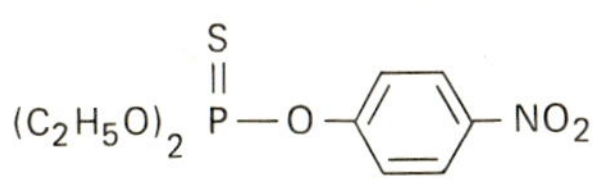

69

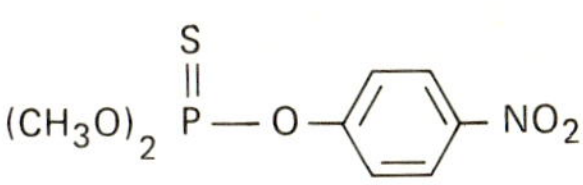

70

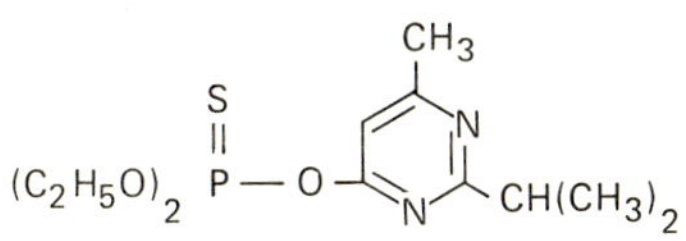

71

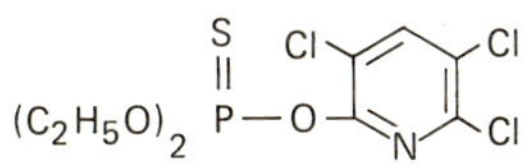

72

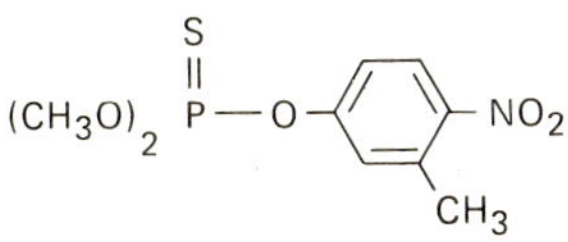

73

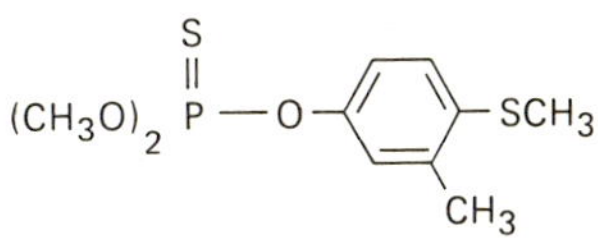

74

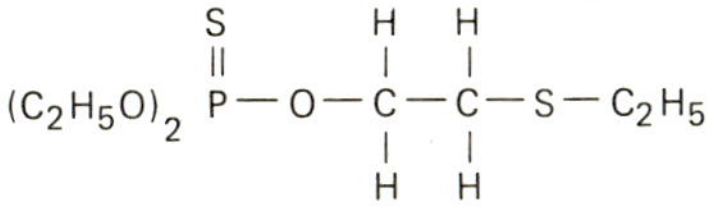

75

2.2.1.3. *Phosphorothiolothionates*

These compounds are usually named phosphorodithioates for simplicity and are considered to be the most commercially important class of phosphorus insecticides. The most widely used insecticides of this class are malathion (76), phenthoate (77), azinophos methyl (78), ethion (79), phorate (80) and dimethoate (81).

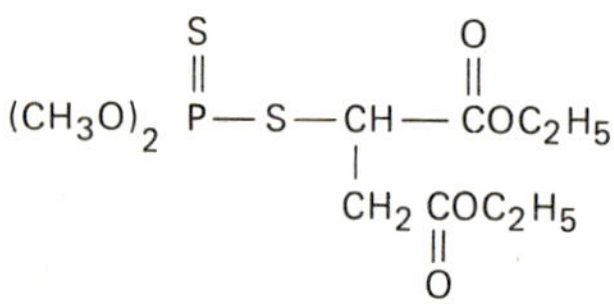

76

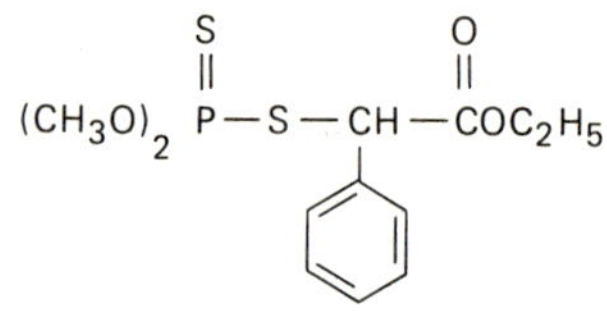

77

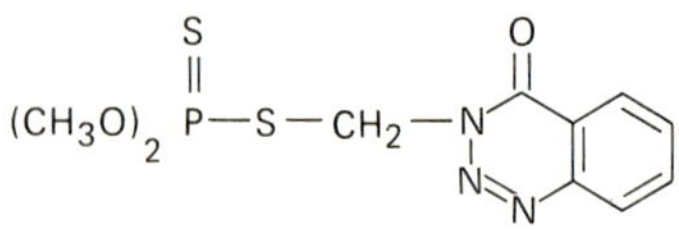

78

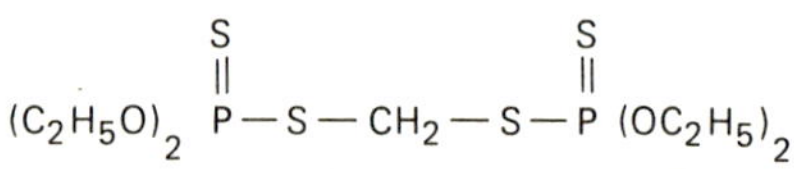

79

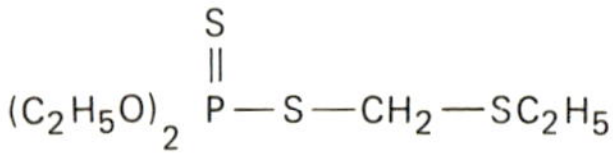

80

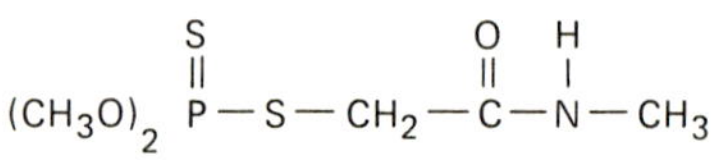

81

2.2.1.4. *Phosphonates and phosphinates*

Some of the commercialized insecticides developed in this class to control soil and plant insects are fonofos (82), EPBP (83), agvitor (84) and leptophos (85).

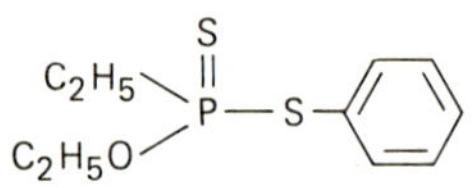

82

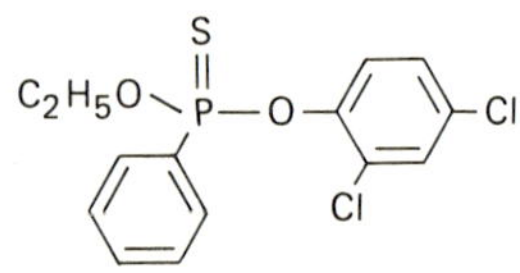

83

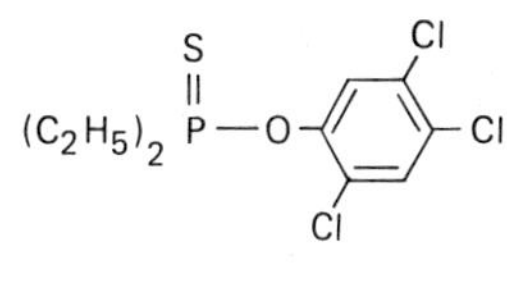

84

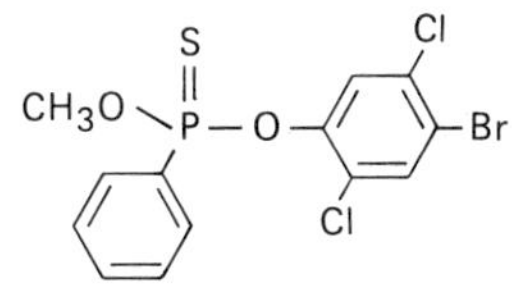

85

2.2.2. Carbamates

The carbamate insecticides are not commonly used against pests in soil. These compounds are closely related to the organophosphorus insecticides in terms of their biological activity. However, their activity is rather more dependent on substituent position and on stereoisomerism than is the case with organophosphorus compounds. The general carbamate structure is:

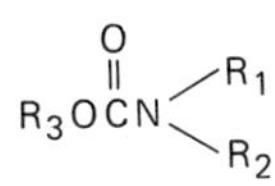

where R_1 and R_2 are hydrogen, methyl, ethyl, propyl or other short chain alkyls, and R_3 is alkyl, phenol, naphthalene, or other cyclic hydrocarbon ring. The commercially used carbamate insecticides that are often used against pests in soils can be divided into three groups. The most commercially useful compounds are *N*-methylcarbamates, which comprise the bulk of carbamate insecticide chemicals. The compounds that may reach the soil are carbaryl (86), methiocarb (87), aldicarb (88) and methomyl (89). Relatively new systemic insecticides include heterocyclic *N*-methylcarbamates, the most widely used of which is carbofuran (90).

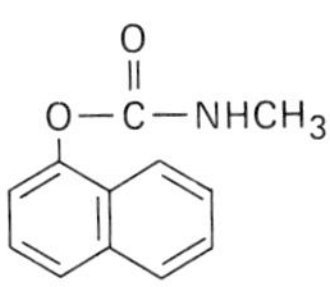

86

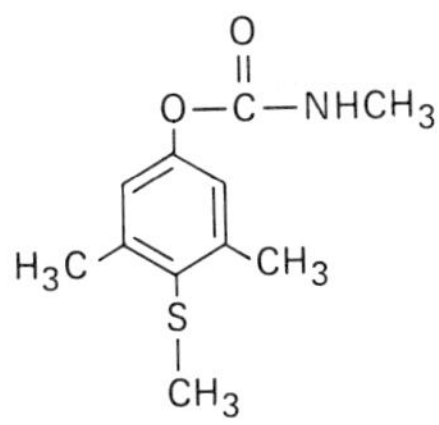

87

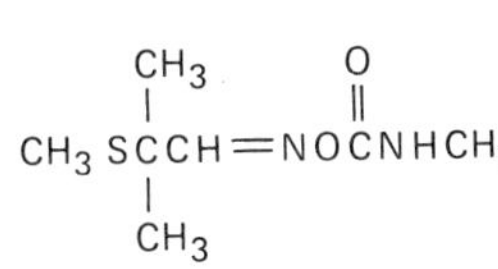

88

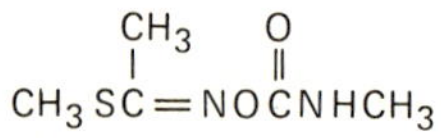

89

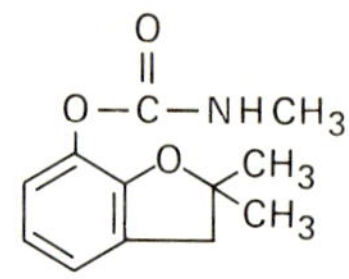

90

2.2.3. Organochlorines

These insecticides are characterized by three major kinds of chemicals: DDT analogues, benzene hexachloride (BHC) isomers, and cyclodiene compounds. They are broad spectrum insecticides active against a great variety of pests.

2.2.3.1. DDT and analogues

DDT (91) has a very wide spectrum of activity among different families of insects and related organisms. It is considered to be one of the most important insecticides ever to appear on the

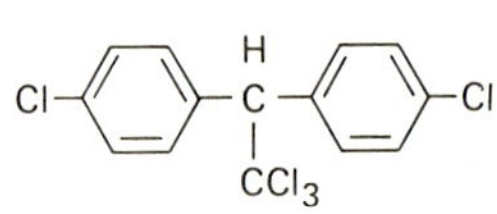

91

market and small traces of this compound can be found in almost all compartments of ecosystems. Methoxychlor (92) is another important DDT analogue.

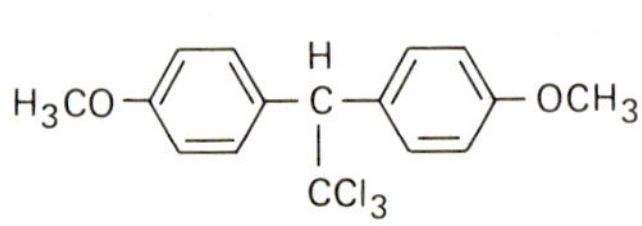

92

2.2.3.2. Benzene hexachloride

The fumigant action of γ-1,2,3,4,5,6-hexachlorocyclohexane (93) also called γ-benzene hexachloride or lindane, makes the compound a useful insecticide. Several structural isomers are possible but the γ-isomer has insecticidal activity.

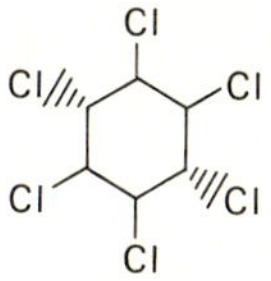

93

2.2.3.3. *Cyclodiene compounds*

During the last fifteen years the use of cyclodiene insecticides has been restricted because of their high mammalian toxicities and extreme persistence in the environment. These compounds are the collective group of synthetic cyclic hydrocarbons. Chlordane (94), aldrin (95), dieldrin (96) and heptachlor (97) are the most powerful general insecticides. They are particularly effective where contact action and long persistence is required.

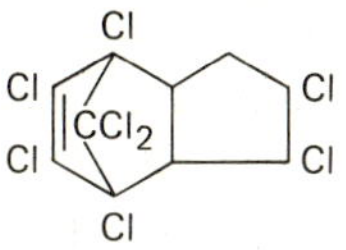

94

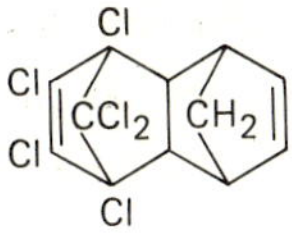

95

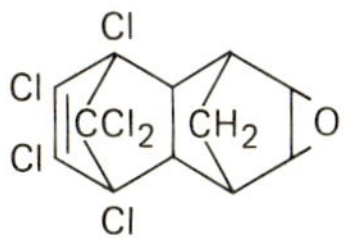

96

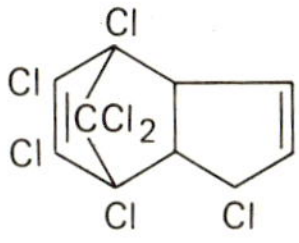

97

2.2.4. Synthetic Pyrethroids

These compounds are readily degraded in soil and have no detectable ill effects on soil microflora and microfauna. They possess high insecticidal activity and low mammalian toxicity. Permethrin (98) is used against a number of insect species of plants and animals in the field. It is stable in air and light and exerts a prolonged residual action. Other important commercially produced synthetic pyrethroids include S-5439 (99) and cypermethrin (100).

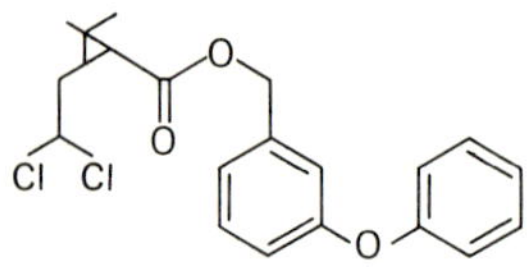

98

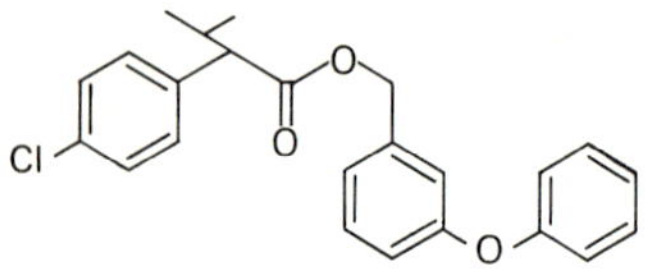

99

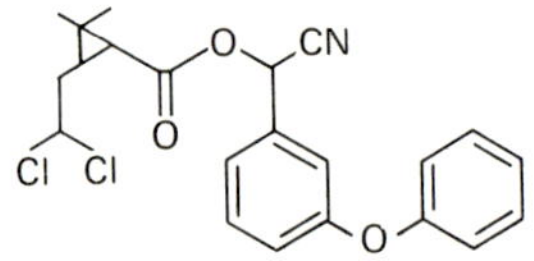

100

2.3. FUNGICIDES

Fungicides are used for crops that lack natural resistance to the fungal species involved. These chemicals are used to treat foliage diseases of some crops, seeds for damping off, soil in seedbeds for root rot, and to control turf and transplant diseases. Some of the fungicides used as seed protectants or for treatment of the soil zone around the seed include hexachlorobenzene (101), chloranil (102), DEXON (103), thiram (104), captain (105), and organic mercurics such as methyl mercury dicyandiamide (106), and phenylmercuric acetate (107). Many of the protective fungicides used in agriculture consist of inorganic compounds of copper, zinc, chromium, nickel and mercury, and organic compounds of tin.

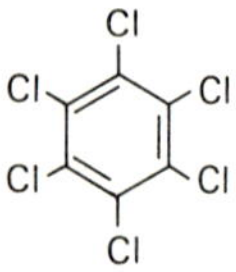

101

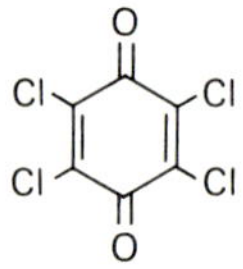

102

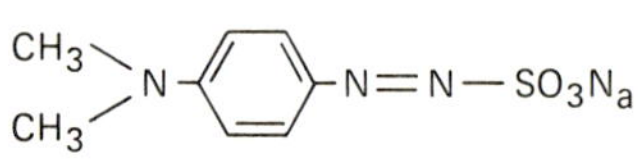

103

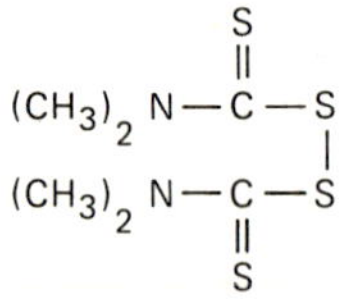

104

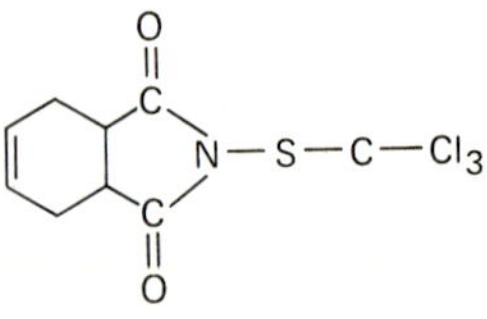

105

$CH_3HgNHC(=NH)NHCN$

106

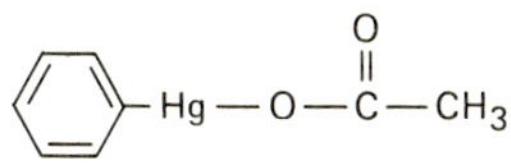

107

Fungus disease has also been controlled by applying systmic fungicides. Some of the synthetic products in commercial use include chloroneb (108), oxycarboxin (109), benomyl (110), thiabendazole (111) and ethirimol (112).

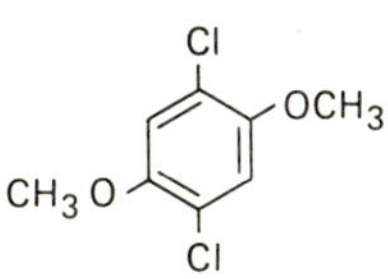

108

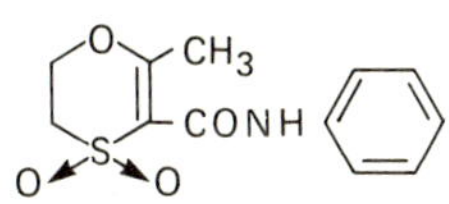

109

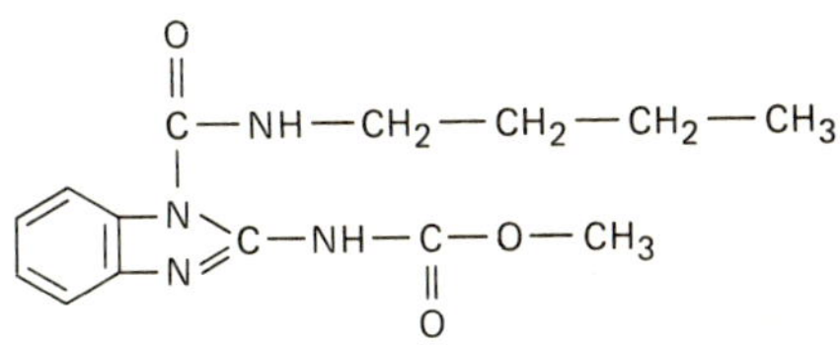

110

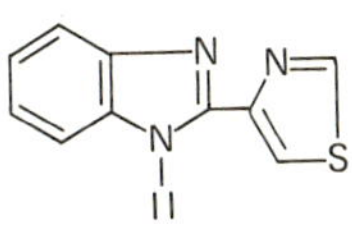

111

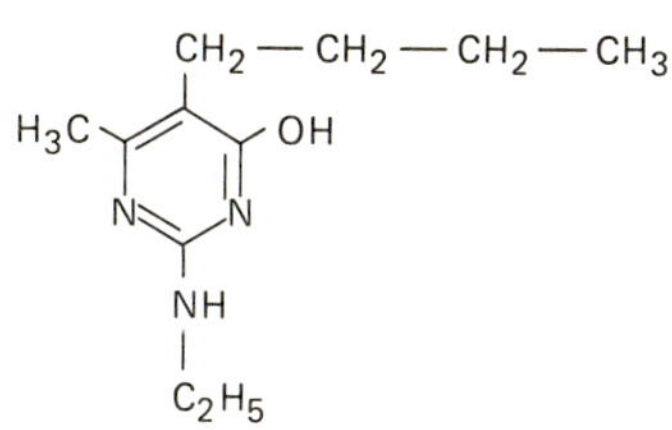

112

2.4. FUMIGANTS

Most of the fumigants are gases at room temperature or liquids and have sufficient volatility to penetrate throughout the upper levels of the soil. Methyl bromide (113), the most volatile fumigant, is almost always applied under the soil cover. Similarly, chloropicrin (114) is also applied under a soil cover. Other commercially available compounds and their uses include formaldehyde

(115) against 'damping off' in surface soil, carbon disulfide (116) against soil fungi, and ethylene dibromide (117), dichloropropene mixture (118) and dibromochloropropane (119) for controlling nematodes in soil.

CH_3Br	CCl_3NO_2	H—CHO	CS_2
113	114	115	116

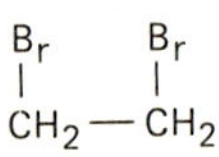

117

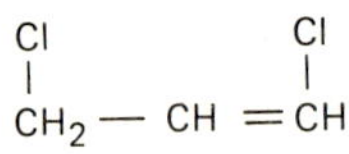

118

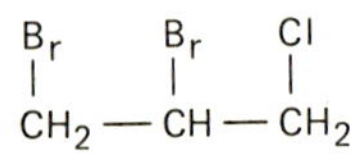

119

REFERENCES

Brooks, G.T., 1974. Chlorinated insecticides, Vol. I, Technology and Application, CRC Press, Cleveland, Ohio, 249 pp.

Crafts, A.S., 1961. The Chemistry and Mode of Action of Herbicides. Interscience Publishers, New York, N.Y., 269 pp.

Eto, M., 1974. Organophosphorus Pesticides: Organic and Biological Chemistry, CRC Press, Cleveland, Ohio, 387 pp.

Khur, R.J. and Dorough, H.W., 1976. Carbamate Insecticides: Chemistry, Biochemistry and Toxicology. CRC Press, Cleveland, Ohio, 301 pp.

Metcalf, R.L., 1971. In: R. White-Stevens (Editor), Pesticides in the Environment, Dekker, New York, N.Y., pp. 1-144.

Chapter 3

PHYSICOCHEMICAL PROCESSES AFFECTING PESTICIDES IN SOIL

The fate of pesticides and their behavior in soil is influenced by several factors including (adsorption, movement and decomposition.) Adsorption, directly or indirectly, influences the magnitude of the effect of other factors. It is considered to be one of the major processes affecting the interactions occurring between pesticides and the solid phase in the soil environment. The main constituents representing the solid phase in soil are clay minerals, organic matter, oxides and hydroxides of aluminum and silicon. A knowledge of the nature of the solid constituents of the soil is essential to understand the adsorption processes. Movement of pesticides in soil can occur by leaching, runoff and volatilization. Information on movement of pesticides is useful in order to predict the probable effectiveness of the chemical. Finally, decomposition processes play an important role in the dissipation of many pesticides in soil. Disappearance of a pesticide from soil can also take place through a number of chemical processes including photodecomposition and chemical reaction or chemical transformation.

This chapter will present a review of the various physicochemical processes that play an important role in influencing the behavior and fate of pesticides in soil. These processes will be discussed under the headings of adsorption, movement, volatilization, chemical conversion and degradation, and photodecomposition.

3.1. ADSORPTION

3.1.1. Characteristics of Soil

The solid phase in soil (mineral and organic) frequently makes up only about 50% of the soil volume, the other half being filled by the soil solution and air. The two major components in soil of significance to adsorption are clay and organic matter.

3.1.1.1. Clay

The term clay is used here to include clay size (<2 μ) crystalline minerals, and crystalline and amorphous oxides and hydroxides. To facilitate an understanding of the adsorption processes, some important features of clay most commonly found in soils are discussed in the subsequent paragraphs. A detailed account of the structure, chemistry and behavior of the clay minerals, oxides and hydroxides is described elsewhere (Grim, 1968; van Olphen, 1963; Greenland, 1965; Marshall, 1967; Bailey and White, 1970; Mortland, 1970; Theng, 1974).

(1) The 1:1 type clay - The kaolinite group is an example of a 1:1 structure (Fig. 3.1a) as it is made up of one sheet of tetrahedrally coordinated cations with one sheet of octahedrally coordinated cations. The surface of the layer on the alumina side is composed of hydroxyls and on the silica side of oxygen. The crystals consist of superimposed unit layers with hydroxyl and oxygen surfaces adjacent to each other (van Olphen, 1963). The thickness of the single layer is about 7.2Å. The 1:1 layer silicate group includes kaolinite, dickite, nacrite, serpentine minerals, and halloysite. Kaolinite particles are relatively large: 0.3 to 4 μm in the maximum dimension and 0.05 to 2 μm thick (Grim, 1968). In general, the 1:1 type layer silicates are electrically neutral or posses a very low negative charge. The surface area and the cation exchange capacity of the kaolinite minerals have relatively low values (Table 3.1).

(2) The 2:1 type clay - The 2:1 clay minerals, such as montmorillonite and vermiculite are made up by combination of two tetrahedrally coordinated sheets of cations, one on either side of an octahedrally coordinated sheet. The thickness of a single 2:1 layer is about 9.6Å. However, the layer height of the minerals depend on the size of the positively charged interlayer group. In the micas or illite, K^+ ions usually balance the charge on the 2:1 layers and the thickness of mica layer is about 10Å (Fig. 3.1c). In the vermiculite, moderately hydrated cations such as Mg^{2+} are found between 2:1 layers and the expansion is restricted to about 4.98Å, the approximate thickness of two molecular layers of water. In the case of montmorillonite, the balancing cations are even more highly hydrated and the layer height depends on the specific nature of the cation and the humidity (Fig. 3.1b). The 2:1

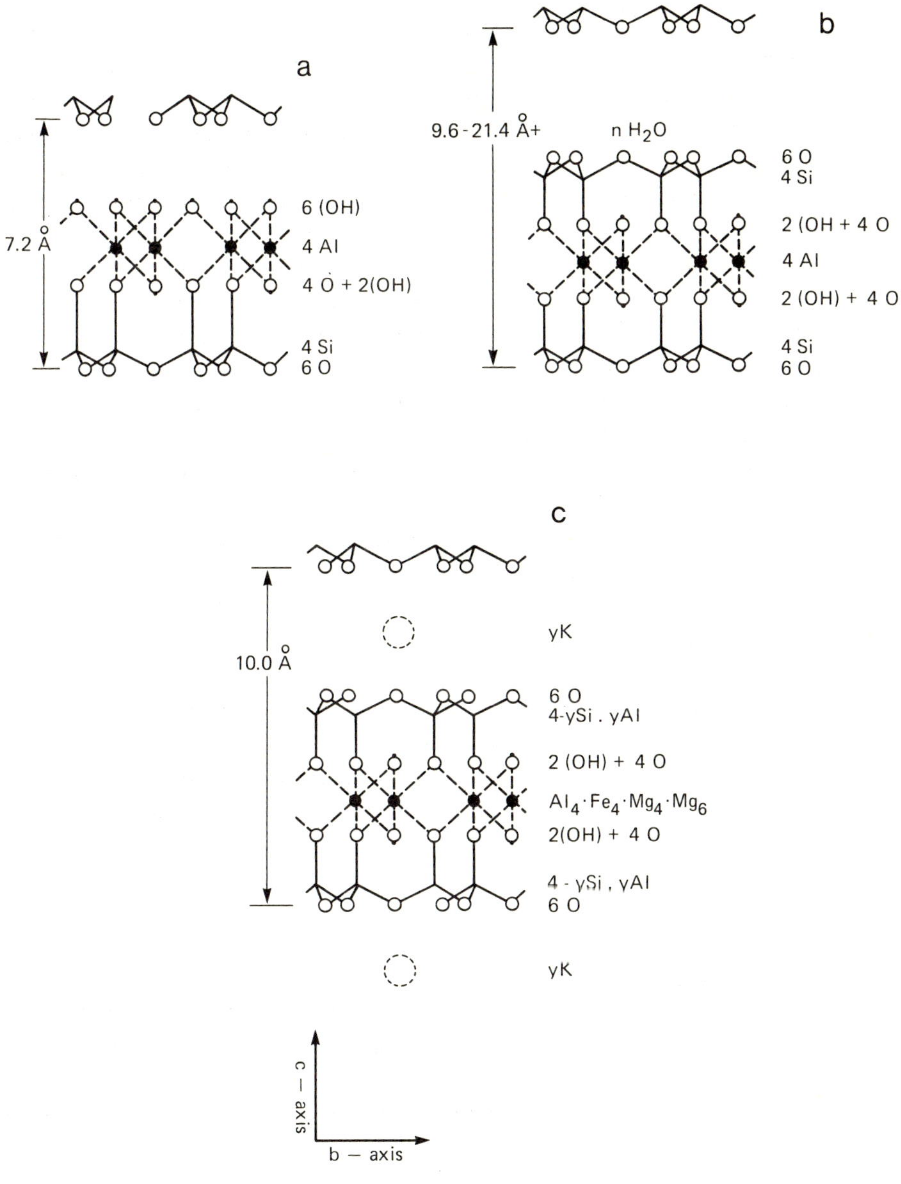

Fig. 3.1. Schematic diagram of the crystal structure of (a) kaolinite, (b) montmorillonite and (c) illite (Toth, 1960).

layer often carries a negative charge due to isomorphous substitution in which Si^{4+} in tetrahedral positions is replaced by Al^{3+} or Mg^{2+} replaces Al^{3+} in octahedral sites. These negative charges are satisfied by exchange cations. The differences in the cation exchange capacity for the crystalline alumino-silicate minerals are due principally to crystalline structure and location of ionic substitution in the lattice. Thus, the expanding 2:1 minerals, such as montmorillonite and vermiculite, have a high cation exchange capacity and high surface area (Table 3.1).

(3) Oxides and hydroxides - Almost all soils contain at least a small proportion of colloidal oxides and hydroxides. The crystalline and amorphous oxides and hydroxide of aluminum, iron and silicon occur in soils as separate phases as well as coatings on surfaces of layer lattice silicates. Some of the amorphous materials such as allophane may have large surface areas and be positively charged whereas some of the crystalline materials may have very low surface areas. Soils containing high amounts of oxides and hydroxides may differ in their adsorptive properties from mineral and organic soil.

3.1.1.2. Organic matter

Soil organic matter plays an important role in affecting the fate of pesticides in the soil environment. It is considered to be one of the most complex materials existing in nature. Organic matter in soil must be chemically characterized if practical

TABLE 3.1

Cation exchange capacity and specific area of clay minerals and humic substances

Soil constituent	Cation exchange capacity (meq/100 g)	Surface area (sq.m./g)
Kaolinite	3 to 15	7 to 30
Illite	10 to 40	65 to 100
Montmorillonite	80 to 150	600 to 800
Vermiculite	100 to 150	600 to 800
Oxides and Hydroxides	2 to 6	100 to 800
Humic Substances	200 to 400	500 to 800

questions regarding its role in affecting pesticides behavior and their fate in soil are to be answered.

Soil organic matter contains compounds that may conveniently be grouped into nonhumic and humic substances. Nonhumic substances include those with definite chemical characteristics such as carbohydrdates, proteins, amino acids, fats, waxes and low molecular weight organic acids. Most of these substances are relatively easily attacked by microorganisms and have a comparatively short life span in soils. Humic substances by contrast, are more stable and constitute the bulk of the organic matter in most soils. They are acidic, dark colored, predominantly aromatic, chemically complex, hydrophilic, polyelectrolyte like materials that range in molecular weights from a few hundred to several thousand.

Based on their solubilities, humic substances are usually partitioned into three main fractions (Fig. 3.2). (1) humic acid (HA), which is soluble in dilute alkali but is precipitated on acidification of the alkaline extract; (2) fulvic acid (FA), which is that humic fraction remaining in solution when the alkaline extract is acidified; that is, it is soluble in both dilute alkali and acid; and (3) humin, which is that humic fraction that cannot be extracted from the soil by dilute base or acid.

From the analytical data published in the literature (Schnitzer

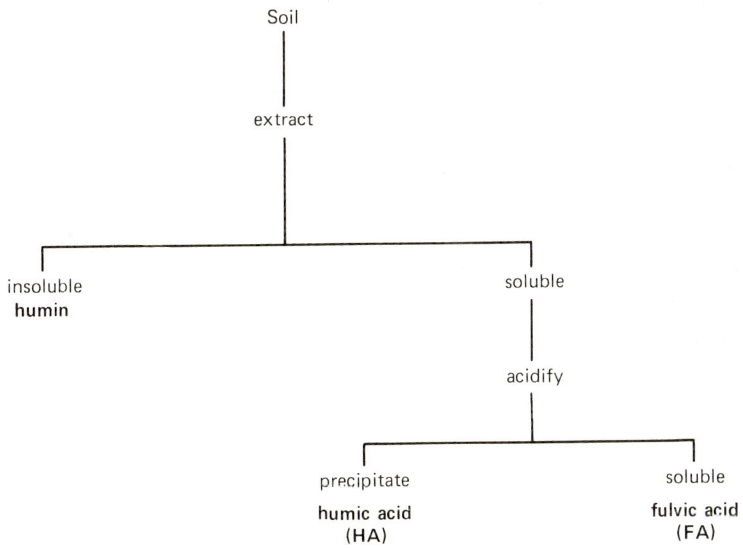

Fig. 3.2. Fractionation of humic substances.

and Khan, 1972) it appears that structurally the three humic fractions are similar, but that they differ in molecular weight, ultimate analysis and functional groups content, with FA having a lower molecular weight but higher content of oxygen containing functional groups per unit weight. The chemical structure and properties of the humin fraction appear to be similar to those of HA. The insolubility of humin seems to arise from it being firmly adsorbed on or bonded to inorganic soil constituents.

Elementary analysis provides information on the distribution of C, H, N, S and O in humic substances. The major oxygen containing functional groups in humic substances are carboxyls, hydroxyls and carbonyls. Some analytical chracteristics of HA and FA are shown in Table 3.2. Elementary and functional group analyses of HA differ from that for FA in the following respect: (1) HA contains more C, H, N and S but less O than does FA; (2) the total acidity and COOH content of FA are approximately twice as great as those of HA; (3) the ratio of COOH to phenolic OH group is about 3 for FA but only approximately 2 for HA; and (4) E_4/E_6 ratios and ESR data also indicate differences between HA and FA (Schnitzer and Khan, 1972). The cation exchange capacity of humic substances is higher than the clay minerals, being of the order of 200 to 400 meq/100 g (Table 3.1).

Generally, humic substances yield uncharacteristic spectra in the ultraviolet (UV) and visible region. Absorption spectra of alkaline and neutral aqueous solutions of HA's and FA's and of acidic, aqueous FA solutions are featureless, showing no maxima or minima; the optical density usually decreases as the wavelength increases. The ratio of optical densities or absorbance of dilute aqueous HA and FA solutions at 465 and 665 nm, usually referred to as E_4/E_6, is widely used for the characterization of these materials. The ratio is independent of concentrations but vary for humic materials extracted from different soil types.

Infrared (IR) spectra of humic materials provide worthwhile information on the distribution of functional groups, and for evaluation effects of different chemical modifications. IR spectrophotometry can be used to ascertain and characterize the formation of metal-humate and clay-humate complexes and to indicate possible interactions of pesticides with humic materials.

Humic substances are known to be rich in stable free radicals which most likely play important roles in polymerization -

TABLE 3.2

Analytical characteristics of humic acid and fulvic acid (Schnitzer and Khan, 1972)

Characteristics	Humic acid	Fulvic acid
Elementary composition (%, on dry ash free basis)		
C	56.4	50.9
H	5.5	3.3
N	4.1	0.7
S	1.1	0.3
O	32.9	44.8
Oxygen containing functional groups (meq/g, on dry ash free basis)		
Total acidity	6.6	12.4
Carboxyl	4.5	9.1
Phenolic hydroxyl	2.1	3.3
Alcoholic hydroxyl	2.8	3.6
Ketonic carbonyl	1.9	2.5
Quinonoid carbonyl	2.5	0.6
Methoxyl	0.3	0.1
E_4/E_6 ratio[1]	4.3	7.1
Free radicals (spin/g x 10^{-18})	0.8	0.2
Line width (G)	3.5	5.0
g value	2.0029	2.0031

[1]Ratio of optical densities of 465 and 665 nm

depolymerization reactions, and in reactions with other organic molecules, including pesticides and toxic pollutants.

Carbohydrates commonly account for 5 to 20% of soil organic matter. Soil carbohydrates are less well understood and a limited information on their origin, composition and behavior is available. Lowe (1978) discussed the significance of soil carbohydrates in relation to environmental problems. Levels and types of carbohydrates present may influence the retention of metal pollutants entering the soil from atmospheric sources or from sewage sludge application. Since microorganisms respond to the levels of readily decomposable substrates like carbohydrates, the latter may indirectly affect the microbial processes that result in the degradation of pesticides in the soil. Organic nitrogen compounds that make up 20 to 50% of the total nitrogen in most surface soils are in bound amino acids and sugars. Less than 1% of the organic nitrogen in soils occurs as purine and pyrimidine bases. Organic phosphorus and sulfur compounds occur in soil primarily as inositol-hexaphosphates and amino acids (e.g. cysteine, cystine, and methionine), respectively.

The presence of organic matter - clay complexes in most of the mineral soils need to be considered in evaluating the importance of soil colloids in pesticide adsorption. It has been observed that up to an organic matter content of about 6%, both mineral and organic surfaces are involved in adsorption (Walker and Crawford, 1968). However, at higher organic matter contents, adsorption will occur mostly on organic surfaces. Stevenson (1976) pointed out that the amount of organic matter required to coat the clay will depend on the soil type and the kind and amount of clay that is present.

For additional information regarding soil organic matter and humic substances, the reader is referred to the books of Schnitzer and Khan (1972, 1978).

3.1.2. Characteristics of Pesticides

A knowledge of a pesticide's structure and some physicochemical properties often permits an estimation of its adsorption behavior. One of the main characteristics of organic pesticides is that most of them are generally low molecular weight compounds with low water solubility. The chemical character, shape and configuration of the pesticide, its acidity or basicity (denoted by pK_a or pK_b), its water solubility, the charge distribution on the cations, the polarity of the molecule, its molecular size and polarizability all affect the adsorption-desorption by soil colloids (Bailey and White, 1970). In the following paragraphs, only those factors that are particularly relevant to pesticide adsorption by soil colloids are discussed briefly.

Four structural factors determine the chemical character of a pesticide molecule and thus influence its adsorption on soil colloids (Bailey and White, 1970).

(1) Nature of functional groups such as carboxyl ($-\overset{O}{\overset{\|}{C}}-OH$), carbonyl (C=O), alcoholic hydroxyl (-OH), and amino ($-NH_2$). The amino groups are specially important as they may protonate, depending on their pK_b and thus adsorb as cations. Both amino and carbonyl groups may participate in hydrogen bonding. In general, adsorption

is characteristically increased with functional groups such as R_3N^+-, -$CONH_2$, -OH, -NHCOR, -NH_2, -OCOR, and -NHR.

(2) Nature of substituting groups that may alter the behavior of functional groups.

(3) Position of substituting groups with respect to the functional groups that may enhance or hinder intramolecular bonding. Position of substituents may permit coordination with transition metal ions.

(4) Presence and magnitude of unsaturation in the molecule that affects the lyophilic-lyophobic balance.

The charge characteristics of a pesticide are probably the most important property governing its adsorption. The charge may be weak, arising from an unequal distribution of electrons producing polarity in the molecule, or it may be relatively strong, resulting from dissociation.

The pH of a system is also an important factor as it governs the ionization of most of the organic molecules. Acidic pesticides are proton donors, which at high pH (one or more pH unit above the pK_a of the acid) become anions due to dissociation. On the other hand basic compounds, when protonated, may behave like organic cations. The adsorption behavior of pesticides that ionize in aqueous solutions to yield cations is different from those that yield anions. Furthermore, nonionic or neutral pesticides behave differently from cationic, basic, or anionic pesticides. Neutral pesticides may be subjected to 'temporary polarization' in the presence of an electrical field, which contributes to adsorption on a charged surface. The availability of mobile electrons, such as π electrons in the benzene ring, influence the polarization of a neutral molecule. Thus, adsorption of neutral pesticides on charged surfaces may increase with molecular size when such increase involves the addition of an aromatic group.

Solubility of a pesticide in water is sometimes considered as an approximate indicator of its adsorption. Bailey et al. (1968) suggested that within a chemical family the magnitude of a pesticide adsorption is directly related to and governed by the degree of water solubility. The hydrophobic character of a pesticide will increase by a decrease in its water solubility thereby resulting in stronger adsorption on soil colloids (Hance, 1965a; Leenheer and Ahlrichs, 1971). An inverse relationship between solubility and adsorption has been observed (Leopold et al., 1960;

Hilton and Yuen, 1963; Ward and Upchurch, 1965). Thus, the adsorption of some acidic herbicides on a muck soil (Weber, 1972), certain nonionic pesticides on organic matter (Carringer et al., 1975), and several substituted ureas on soil (Wolf et al., 1958) was found to be inversely related to the water solubilities of the compounds. On the other hand, no relationship has been found between water solubility of certain pesticides and adsorption on various surfaces (Harris and Warren, 1964; Hance, 1965a, 1967; Weber, 1966, 1970b). Bailey et al. (1968) found a direct relationship between water solubility and adsorbability for some *s*-triazines and substituted ureas on sodium and hydrogen clays. It appears that for a particular family of pesticides, several factors may be interacting in determining direct, inverse or no relationship between water solubility and absorbability.

For detailed information on the nature and characteristics of pesticides the reader is referred to the work of Metcalf (1971) and Melnikov (1971).

3.1.3. Adsorption Isotherms

Adsorption of pesticides is generally evaluated by the use of adsorption isotherms. An isotherm represents a relation between the amount of pesticide adsorbed per unit weight of adsorbent and the pesticide concentration in the solution at equilibrium. Giles et al. (1960) investigated the relation between solute adsorption mechanisms on solid surfaces and the types of adsorption isotherms obtained. They developed an empirical classification of adsorption isotherms into four main classes according to the initial slope (Fig. 3.3). The S-type isotherms are common when the solid has a high affinity for the solvent. The initial direction of curvature showed that adsorption becomes easier as concentration increases. In practice, the S-type isotherm usually appears when the solute molecule is monofunctional, has moderate intermolecular attraction, and meets strong competition for substrate sites from molecules of the solvent or of another adsorbed species. The L-type curves, the normal or Langmiur isotherms, are the best known and represent a relatively high affinity between the solid and solute in the initial stages of the isotherm. As more sites in the substrate are filled, it becomes increasingly difficult for solute molecules to find a vacant site available. The C-type curves are

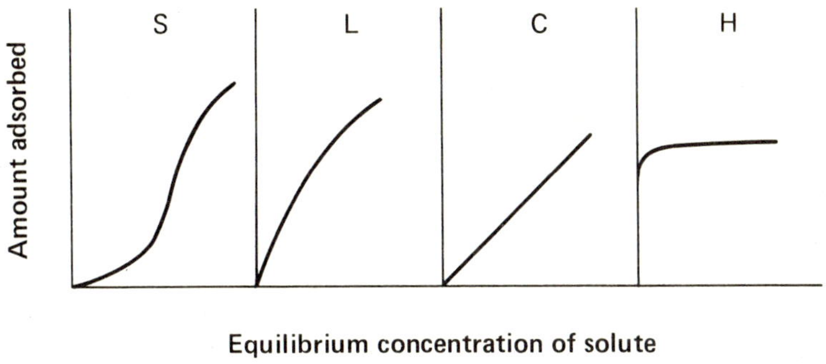

Fig. 3.3. Classification of adsorption isotherms according to Giles et al. (1960). Reproduced from 'Pesticide in Soil and Water', 1974, p. 45, by permission of the Soil Science Society of America.

given by solutes that penetrate into the solid more readily than does the solvent. These curves are characterized by the constant partition of solute between solution and substrate, right up to the maximum possible adsorption, where an abrupt change to horizontal plateau occurs. The H-type curves are quite uncommon and occur only when there is very high affinity between solute and solid. This is a special case of the L-type curves, in which the solute has such high affinity that in dilute solutions it is completely adsorbed, or at least there is no measurable amount remaining in solution. The initial part of the isotherm is therefore vertical. The foregoing four classes of isotherms have been referred to in the literature on many instances concerning pesticide adsorption on soil colloids.

In general, the following two mathematical equations have been used for a quantitative description of pesticide adsorption on soil materials.

(1) Freundlich adsorption equation - The empirically derived Freundlich eq. 3.1 has been used to describe the adsorption of pesticides by soil, organic matter and clay minerals in the majority of published reports. The Freundlich equation can be expressed as:

$$\frac{x}{m} = KC^{\frac{1}{n}} \qquad (3.1)$$

where x/m is the ratio of pesticide to the adsorbent mass, C is the pesticide concentration in solution upon achieving equilibrium, and K and n are constants. The form 1/n emphasizes that C is raised to a power less than unity. When eq. 3.1 is expressed in the logarithmic form, a linear relationship is obtained:

$$\log \frac{x}{m} \quad \log K + \frac{1}{n} \log C \qquad (3.2)$$

Normally, within a reasonable range of pesticide concentration, the relationship between log x/m and log C is linear, with 1/n being constant. In comparing adsorptivity of various pesticides by different surfaces, the K value may be considered to be a useful index for classifying the degree of adsorption. The necessary conditions are that 1/n values be approximately equal and determination be made at the same C value (Hance, 1967). In general, K and 1/n values for the adsorption of pesticides on soil organic matter or clay minerals decrease and increase, respectively, with increase in temperature (Haque and Sexton, 1968; Khan, 1973b, 1974b, 1977). Grover (1971, 1977) reported that the Freundlich K values as calculated above were similar to K_d values (μg/g adsorbed divided by μg/ml in solution at equilibrium).

The Freundlich isotherms for phorate sorption on soils are shown in Fig. 3.4 (Felsot and Dahm, 1979). Similar isotherms have been reported for various other pesticides on different surfaces. In an ideal situation, the slope of the isotherm would equal one, and there would be unlimited adsorption as the equilibrium concentration continually increased. The isotherms (Fig. 3.4) had slopes ranging from 0.80 to 0.99, which are consistent with the values reported for other pesticides (Hamaker and Thompson, 1972). The variable slopes obtained for the different pesticides-soil systems indicate that sorption in soil is a complex phenomenon involving different types of adsorption sites with different surface energies.

Felsot and Dahm (1979) recently investigated the adsorption of organophosphorous and carbamate insecticides by different soils. They determined the relationship among log K values for adsorption, soil variables and pesticide physicochemical characteristics (Table 3.3). Significant correlations were found among log K, log organic matter, and cation exchange capacity. Furthermore, a significant correlation between log inverse water solubility and

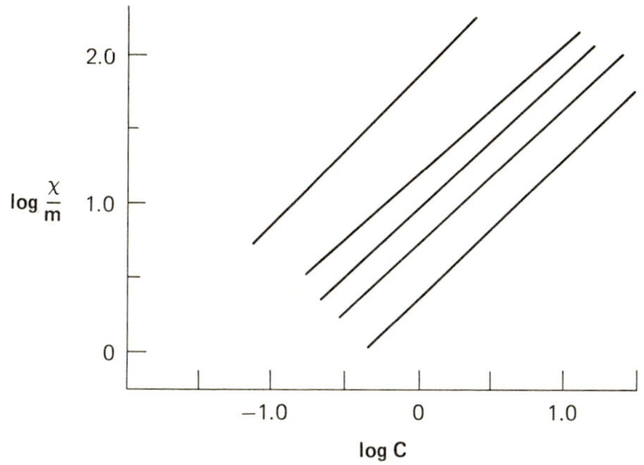

Fig. 3.4. Freundlich isotherms for phorate sorption of five soils (Felsot and Dahm, 1979).

log partition coefficient (PC) was observed. A similar relationship was reported by Chiou et al. (1977) for a number of insecticides. The partition coefficient of a pesticide indicates its tendency to favor a nonpolar milieu (e.g., octanol or other hydrophobic molecules and surfaces) over a polar one (e.g. water or clay surface) and it may be defined as PC = concentration in octanol/ concentration in water. A significant correlation was found among water solubility, partition coefficient and parachor (Table 3.3).

TABLE 3.3

Correlation coefficients among log K values for adsorption, soil variables and insecticides physicochemical characteristics (Felsot and Dahm, 1979)

	log K	log OM	CEC	Clay	pH	log $(S)^{-1}$	log PC
log OM	0.469**						
CEC	0.439**	0.960*					
Clay	0.393	0.718*	0.618*				
pH	-0.043	-0.096	0.118	-0.288			
log $(S)^{-1}$	0.799*	-0.077	-0.086	-0.019	-0.016		
log PC	0.794*	-0.010	-0.111	-0.025	-0.020	0.978*	
Parachor	0.765*	-0.068	-0.076	-0.017	-0.014	0.989*	0.954*

OM = organic matter, CEC = cation exchange capacity, S = solubility, PC = partition coefficient, * = significant at 1% level, and ** = significant at 5% level

Parachor is an approximate measure of the molar volume of a molecule and is a constitutive and additive function of molecular structure (Lambert, 1967).

(2) Langmuir adsorption equation - The Langmuir adsorption equation was initially derived from the adsorption of gases by solids using the following assumptions: (i) the energy of adsorption is constant and independent of surface charge; (ii) adsorption is on localized sites and there is no interaction between adsorbate molecules; and (iii) the maximum adsorption possible is that of a complete monolayer. The Langmuir adsorption equation may be expressed in terms of concentration in the form:

$$\frac{x}{m} = \frac{K_1 K_2 C}{1+K_1 C} \tag{3.3}$$

The terms x/m and C have been defined earlier, K_1 is a constant for the system dependent on temperature and K_2 is the monolayer capacity. The reciprocal of eq. 3.3 gives:

$$\frac{1}{x/m} = \frac{1}{K_2} + \frac{1}{K_1 K_2 C} \tag{3.4}$$

A plot of 1/(x/m) against 1/C, should give a straight line with an intercept of $1/K_2$ and a slope of $1/(K_1K_2)$ when the Langmuir relation holds. The adsorption of a number of pesticides on various soil surfaces was found to conform to an isotherm type which was similar to Langmuir model for adsorption (Weber and Gould, 1966; Li and Felbeck, 1972a; Karickhoff and Brown, 1978; Juo and Oginni, 1978).

Singhal and Singh (1976) observed that the adsorption of nemagon on montmorillonite suspension yielded H-type isotherms. Their data agreed with the Langmuir equation (3.4). Fig. 3.5 shows the adsorption of nemagon on H-montmorillonite.

Under certain conditions both the Freundlich and Langmuir equations may reduce to linear relationship. In the case of the Freundlich eq. 3.1, if the exponent 1/n is 1, the adsorption will be linearly proportional to the solution concentration. It has been generally found, in practice, that adsorption of pesticides on soil surfaces do fit the Freundlich equation with an exponent

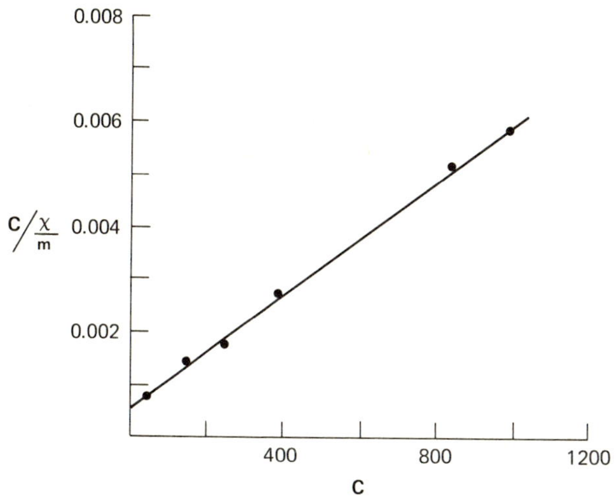

Fig. 3.5. Langmuir isotherm for nemagon adsorption on H^+-montmorillonite (Singhal and Singh, 1976).

close to unity. In the case of the Langmuir eq. 3.3, the denominator $1 + K_1C$ becomes indistinguishable from 1 at low concentration. Thus, the amount adsorbed becomes directly proportional to the concentration in solution.

Eqs. 3.1 and 3.3 will not be obeyed if the adsorption of pesticides is predominantly due to an ion exchange mechanism. Burns et al. (1973a) examined the validities of two ion exchange isotherm equations for the adsorption of paraquat cation (P^{2+}) on a hydrogen saturated HA. The Rothmund-Kornfeld equation is given by Burns et al. (1973a):

$$\frac{[\bar{P}^{2+}]}{[\bar{H}]^2} = K \left(\frac{[P^{2+}]}{[H]^2}\right)^{\frac{1}{n}} \tag{3.5}$$

where the superimposed bars refer to the ions in the adsorbent. Eq. 3.5 is reduced to an expression of the law of mass action when n = 1. The logarithmic form of eq. 3.5 can be expressed as:

$$A = \log K + \left(\frac{1}{n}\right)S \tag{3.6}$$

where $A = \log [\bar{P}^{2+}] - 2 \log [\bar{H}^{+}]$ and $S = \log [P^{2+}] - 2 \log [H^{+}]$. This can be used to test the data in both Rothmund-Kornfeld and mass action equations. Burns et al. (1973a) found that only the Rothmund-Kornfeld eq. 3.5 satisfactorily fitted the results. However, at low concentrations small deviations were observed, which were attributed to non exchange adsorption because of deviations from Donnan behavior at low concentrations. Neither Freundlich nor Langmuir plots fitted the data, although some of the data at lower concentration levels were in reasonable accord with the Freundlich model for adsorption.

According to Burns and Hayes (1974), it is possible to distinguish between ionic and other mechanisms of adsorption by using the isotherm equation. Thus, carefully controlled adsorption studies at different temperatures can give some idea of the mechanism involved.

3.1.4. Mechanisms of Adsorption

Several mechanisms have been proposed for adsorption of pesticides by soil constituents. Two or more mechanisms may occur simultaneously depending upon the nature of the pesticide and soil surface. The mechanisms most likely involved in the adsorption of pesticides on soil colloids are outlined below.

(1) Van der Waals attractions - Van der Waals forces are involved in the adsorption of nonionic, nonpolar molecules or portions of molecules. Van der Waals forces result from short range dipole-dipole interactions of several kinds. The additive nature of Van der Waals forces between the atoms of adsorbate and adsorbent may result in considerable attraction for large molecules. Haque and Coshow (1971) attributed adsorption of isocil on both montmorillonite and kaolinite to Van der Waals interactions. The adsorption of carbaryl and parathion on soil organic matter in aqueous systems is considered to be physical involving Van der Waals bonds between the hydrophobic portions of the adsorbate molecules and the adsorbent surface (Leenheer and Ahlrichs, 1971). Nearpass (1976) suggested that the principal adsorption mechanism for picloram by humic materials was molecular adsorption due to Van der Waals forces.

(2) Hydrophobic bonding - Nonpolar pesticides or compounds whose molecules often have nonpolar regions of significant size

in proportion to polar regions are likely to adsorb onto the hydrophobic regions of soil organic matter. Water molecules present in the system will not compete with nonpolar molecules for adsorption on hydrophobic surfaces. The potential importance of the hydrophobic fractions of organic matter for the retention of pesticides was cited by Hance (1969b). This type of bonding may also be largely responsible for the strong adsorption by soil organic matter of many pesticides such as DDT and other organochlorine insecticides. Lipids in the organic matter are the primary sites for adsorption of chlorinated hydrocarbon pesticides. As much as 20% lipid content is not uncommon for some peat and muck soils (Stevenson, 1966). Lipids are also associated with soil humus (Khan and Schnitzer, 1972; Schnitzer and Khan, 1972). Thus, association of nonpolar (chlorinated hydrocarbons) pesticides with the lipid fraction of soil organic matter and humus might be described by hydrophobic bonding (Pierce et al., 1971). This also explains the relative independence of pesticide adsorption on moisture in soils with high organic content. Nonpolar portions of the humic polymer and hydrophobic molecules trapped within the polymer could also provide hydrophobic binding sites for DDT (Pierce et al., 1974). The hydrophobic portion of peats such as fats, waxes and resins can be a significant adsorbent of phenylureas (Hance, 1969b; Morita, 1976). The adsorption of pesticides involving this mechanism would be independent of pH (Hance, 1965b; Walker and Crawford, 1968). Methylation of organic matter or humic substances to block hydrophilic hydroxyl groups would increase the adsorption by this mechanism. In view of this concept, adsorption of pesticides by a soil can be considered to be primarily a matter of partitioning between organic matter and water (Lambert et al., 1965; Lambert, 1968).

(3) Hydrogen bonding - This is a special kind of dipole-dipole interaction in which the hydrogen atom serves as a bridge between two electronegative atoms, one being held by covalent bond and the other by electrostatic forces. There is a parallel between hydrogen bonding and protonation (Hadzi et al., 1968). Protonation may be considered as a full charge transfer from the base (electron donor) to the acid (electron acceptor). The hydrogen bonding interaction is a partial charge transfer. Hydrogen bonding appears to be the most important mechanism for adsorption of polar nonionic organic molecules on clay minerals.

The presence of oxygen containing functional groups, as well as amino groups, on organic matter indicates that adsorption could occur by the formation of a hydrogen bond with organic pesticides containing similar groups (Khan and Schnitzer, 1971; Khan, 1974e, 1977b). For example, carbonyl oxygens on pesticide molecules may bound to amino hydrogens or hydroxyl groups on the organic matter. Additional sites for hydrogen bonding by soil organic matter includes -SH and -O- linkages (Stevenson, 1972). Hayes (1970) stressed the participation of a hydrogen bonding mechanism in *s*-triazines and organic matter interactions. Evidence for this type of bonding was obtained from infrared studies by Sullivan and Felbeck (1968). They observed that hydrogen bonding may take place between C=O groups of the humic compounds and the secondary amino groups of *s*-triazines. The heat of HA-atrazine complex formation was estimated as 8-13 Kcal/mole, which most likely is the heat of formation of one or more hydrogen bonds. Binding of *s*-triazines, such as simazine, by hydrogen bonds with weakly acidic groups of HA may result in the formation of a stable complex (Maslennikova and Kruglow, 1975).

Anionic pesticide adsorption at pH values below their pK_a values can be attributed to adsorption of the unionized form of the molecule on organic surfaces. Thus, hydrogen bonding may take place between the COOH group and C=O or NH group of organic matter (Kemp et al., 1969). Hydrogen bonding would be limited to acid conditions where COOH groups are unionized (Stevenson, 1972).

Several hydrogen bonds utilizing oxygen atoms on the clay surface or edge hydroxyls may bind organic pesticides to clay minerals (Bailey et al., 1968). The hydrogen bond associated with the 'water bridge' between the exchange cation and a polar organic cation plays an important role in the binding of organics on clays under normal soil conditions. The binding of dasanit® on both Na- and H-montmorillonite in clay-water suspension may be attributed to hydrogen bonding by water bridging (Bowman, 1973). Malathion is adsorbed on each homoionic clay saturated with Na^+, Ca^{2+}, Cu^{2+}, Fe^{3+}, or Al^{3+} by hydrogen bonding between the carbonyl oxygen atoms and hydration water shells of the cation (Bowman et al., 1970). The adsorption of 2,4-D acid on montmorillonite may involve hydrogen bonding of the C=O group to the hydroxyls of the clay surface (Dieguez-Carbonell and Pascual, 1975).

(4) Charge transfer - In the formation of charge transfer complexes, electrostatic attraction takes place when electrons are transferred from an electron rich donor to an electron deficient acceptor. Charge transfer interaction will take place only within short distances of separation between the interacting species. The formation of charge transfer complexes has been postulated as the possible mechanism involved in the adsorption of *s*-triazines onto soil organic matter and clay minerals (Hayes, 1970; Haque et al., 1970). The charge transfer reactions are particularly important in explaining the high adsorption of methylthiotriazines onto organic matter (Hayes, 1970). Burns et al. (1973b) postulated the involvement of charge transfer mechanisms in paraquat adsorption by HA. However, their study involving an ultraviolet spectroscopic technique failed to provide evidence for such a mechanism in the formation of the paraquat-HA complex in an aqueous system. Presumably, the ultraviolet methods are not sufficiently sensitive to detect any charge transfer interactions. Khan (1973b, 1974a,e) provided evidence for such interactions using infrared spectrophotometry. The interaction of bipyridylium herbicides with humic materials resulted in a shift of C-H out-of-plane bending vibrations from 815 to 825 cm^{-1} for paraquat, and from 729 to 765 cm^{-1} for diquat (Fig. 3.6). The observed shifts in the out-of-plane C-H vibration frequencies provide evidence for the charge transfer complex formation between the humic materials and the bipyridylium herbicides.

In a similar study, Haque et al. (1970) reported marked changes in the out-of-plane C-H vibration frequencies in the infrared spectra of diquat- and paraquat-montmorillonite complexes. For the paraquat and diquat complexes this band shifted from 854 to 834 cm^{-1} and from 793 to 782 cm^{-1}, respectively. They concluded that the shifts resulted from the organocation-anionic clay surface associations through charge transfer processes. The data presented by Burdon et al. (1977) supported this view by showing that positive charges in the bipyridylium cations are distributed around the molecules and are greatest in the positions ortho and para to the heterocyclic nitrogen atoms. Their x-ray data demonstrated close contact between the bipyridylium cations and the interlamellar surfaces of montmorillonite. If it is assumed that the negative charges on the clay are not point charges and that these charges are to some extent smeared along the clay surface,

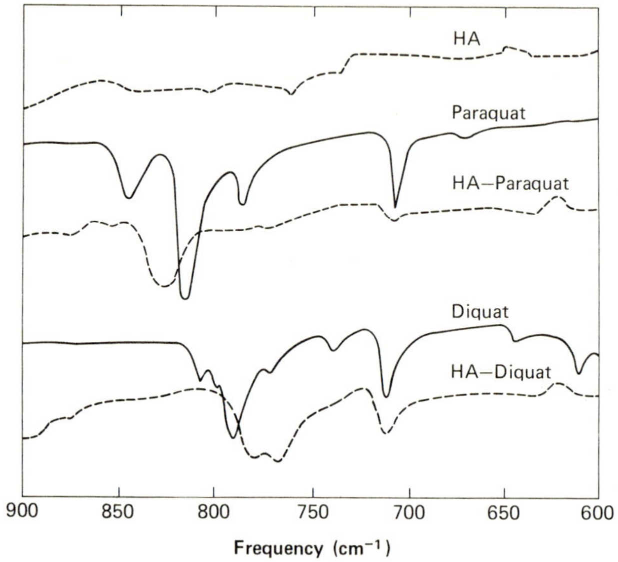

Fig. 3.6. Infrared spectra of humic acid (HA), herbicide and HA-herbicide complex in the region 600-900 cm^{-1} on expanded scale (Khan, 1974a). Published by permission of the American Society of Agronomy, Crop Science Society of America, and the Soil Science Society of America.

then only is it plausible that charge transfer processes are involved in the clay-bipyridylium cation interactions (Burdon et al., 1977).

(5) Ion exchange - Ion exchange adsorption takes place for those pesticides that either exist as cations or that become positively charged through protonation. Adsorption of cationic pesticides, such as paraquat and diquat, via cation exchange functions through COOH and phenolic-OH groups associated with the organic matter (Broadbent and Bradford, 1952; Schnitzer and Khan, 1972). The adsorption is always accompanied by the release of a significant concentration of hydrogen ions (Best et al., 1972; Khan, 1974a). According to Stevenson (1976), diquat and paraquat can react with more than one negatively charged site on soil humic colloids, such as through two COO^- ions, a COO^- ion plus a phenolate ion combination, or a COO^- ion (or phenolate ion) plus a free radical site. Due to the ionic character of diquat and paraquat, these compounds are also readily adsorbed on clay

minerals. The importance of ion exchange to the adsorption of these compounds is reflected by the greater adsorption of paraquat on montmorillonite at high pH and the less adsorption on kaolinite (Weber et al., 1965). These cationic pesticides readily replace inorganic cations on montmorillonite and are adsorbed to the extent of the cation exchange capacity (Weed and Weber, 1969). Paraquat and diquat are difficult to remove from montmorillonite by ion exchange with inorganic cations, but are displaced more easily from kaolinite and vermiculite.

Burns et al. (1973b) and Khan (1974a) utilized IR spectroscopy to demonstrate that ion exchange is the predominant mechanism for adsorption of bipyridylium herbicides by humic substances. Spectra for the HA herbicide complexes are presented in Fig. 3.7. It can be seen that upon addition of herbicides the intensity of the 1720 cm^{-1} band (carbonyl of carboxylic acid) diminished while that at 1610 cm^{-1} (carboxylate) increased. This indicated a conversion of COOH to COO^- groups, which react with bipyridylium cations to form carboxylate bonds. Notice that the 1720 cm^{-1} band did not disappear completely, indicating that a considerable proportion of H^+ in COOH remained inaccessible to the large herbicide cations. HA and FA retains paraquat and diquat in amounts that are considerably less than the exchange capacity of humic materials (Khan, 1973b). The large size of the organic cations seems to result in steric hindrance so that they may not be exchanged with ionizable H^+ as effectively as the smaller inorganic cations.

Further evidence for the ion exchange mechanism was procured by the potentiometric titrations of HA and pesticide-HA complexes (Fig. 3.8). The decreases in consumption of alkali for the pesticide-HA complexes titration (curves b, c vs curve a) suggest that ionization of acid functional groups are involved in the bipyridylium cations interactions with humic materials (Khan, 1974a).

It was suggested earlier that charge transfer mechanisms are also involved in the adsorption of bipyridylium cations by HA. An estimate of the relative importance of charge transfer and ion exchange mechanisms in the adsorption of bipyridylium cations by HA will remain a matter of conjecture until more information is available. However, judging from the data available in the literature it appears that an ion exchange mechanism plays a dominant role in the adsorption processes.

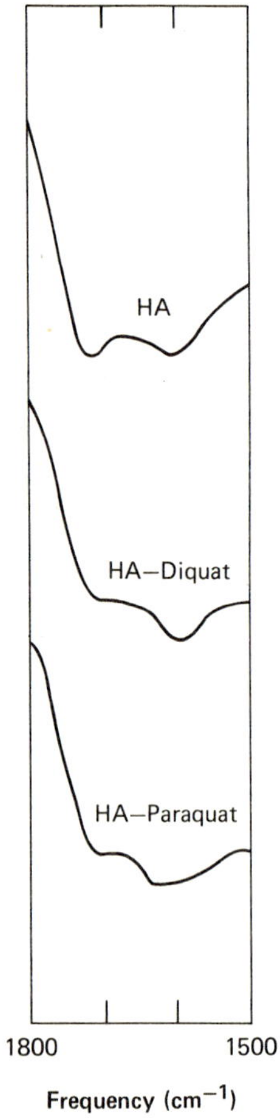

Fig. 3.7. Infrared spectra of humic acid (HA) and HA-herbicide complex in the region 1500-1800 cm^{-1} (Khan, 1974a). Published by permission of the American Society of Agronomy, Crop Science Society of America, and the Soil Science Society of America.

The cationic adsorption mechanism is also responsible for the adsorption of less basic pesticides, such as *s*-triazines on organic matter and clay minerals (Weber et al., 1969; Gaillardon, 1975). The pesticide may become cationic through protonation, either in the soil solution or during adsorption. Thus, a weakly basic pesticide may be protonated and adsorbed on soil colloids according to the following series of equations:

$$P + H^+ = PH^+ \qquad (3.7)$$

where P = weakly basic organic pesticide. When the solution pH is equal to the pK_a of the compound, 50% of the basic pesticide molecules are protonated. In this case, the pK_a is derived from the expression:

$$K_a = \frac{[H^+][P]}{[PH^+]} \quad (3.8)$$

Maximum adsorption of *s*-triazines by soil colloids occurs at pH levels near the pK_a of the respective compound (Weber et al., 1969). Thus, the adsorption capacity of organic matter, humic substances and clay minerals for *s*-triazines follow the order expected on the basis of pK_a values for the compounds (Weber et al., 1969; Weber, 1970a; Gilmour and Coleman, 1971). The pH of the soil

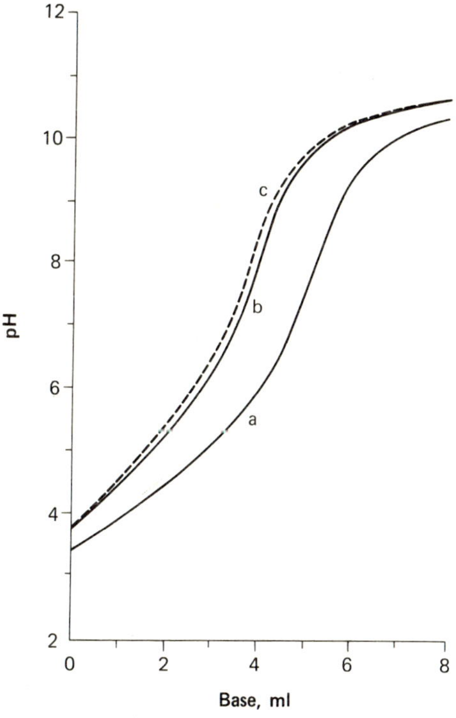

Fig. 3.8. Potentiometric titration curves of (a) humic acid (HA), (b) HA-paraquat complex and (c) HA-diquat complex (Khan, 1974a). Published by permission of the American Society of Agronomy, Crop Science Society of America, and the Soil Science Society of America.

solution will govern the ionization of the acidic functional groups on organic matter that may be available for cation exchange. This would also affect the adsorption of weakly basic pesticides (Nearpass, 1965; 1969; 1971). Reduction in solution pH results in an increase in the protonated species. For the subsequent adsorption of PH^+, it should compete with initially adsorbed cation (M+).

$$PH^+ + MR \rightleftharpoons M^+ + PHR \tag{3.9}$$

where R is the soil cation exchanger. Sullivan and Feldbeck (1968) showed that ion exchange could take place between a protonated secondary amine group on *s*-triazines and a carboxylate anion on the HA. Gilmour and Coleman (1971) also suggested an ion exchange process between protonated *s*-triazine and Ca-humate. Larger Ca-saturation of HA resulted in less *s*-triazine adsorption. Adsorption was greater for more strongly basic *s*-triazines as compared to weakly basic *s*-triazines under the same conditions because, at a given pH, the proportion of *s*-triazine was greater.

Protonated hydroxyatrazine has been shown to be adsorbed as an organic cation at the surface of the H^+- and Al^{3+}-montmorillonite (Russell et al., 1968a,b). Propazine was also protonated and hydrolyzed in the presence of H^+-montmorillonite (Cruz et al., 1968). The adsorption of amitrole by montmorillonite occurred after protonation of the compound by the highly polarized water molecules in direct coordination with the cations on the exchange sites of montmorillonite (Russell et al., 1968a,b).

Protonation may also occur by H^+ already countering the charge on R^- and the protonated pesticide remains on the surface as counter ion:

$$P + HR \rightleftharpoons PHR \tag{3.10}$$

Thus, the acidity of the soil colloid surface will influence the protonation of the adsorbed basic pesticide molecule. The pH at the surface of soil colloids may be as much as two pH units lower than that of the liquid environment (Hayes, 1970). Thus, the protonation of a basic pesticide may occur even though the measured pH of the water-adsorbent system is greater than the pK_a of the compound.

Adsorption of some benzimidazole fungicides on clay surfaces has been attributed to protonation of the basic organic molecule (Aharonson and Kafkafi, 1975a). Thus, the pH dependence of the adsorption of benzimidazole derivatives such as, 2-benzimidazole-carbamic acid methyl ester (120) and thiobendazole (122) by soils may also be due to protonation of the molecules on the soil surface (Aharonson and Kafkafi, 1975b) (*Scheme 3.1*).

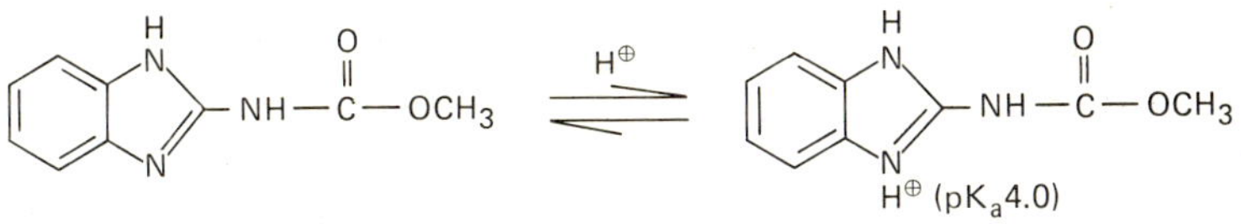

120 121

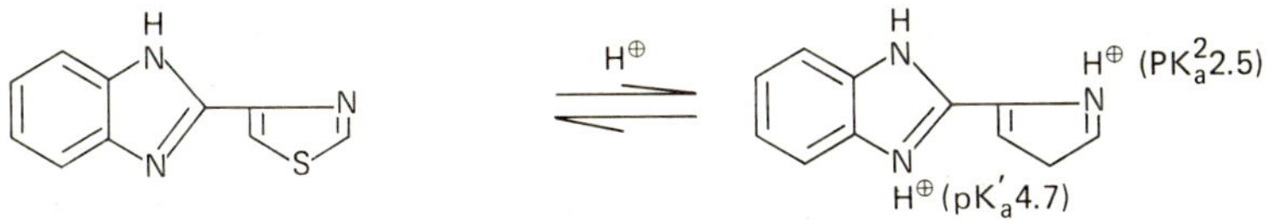

122 123

Scheme 3.1

Diprotonation of picloram at pH values below 1 was reported by Nearpass (1976). The cation thus formed cannot compete with H^+ for adsorption sites, thereby resulting in a slight decrease in picloram adsorption in this pH region.

Ion exchange adsorption of pesticides by soil colloidal constitutents will also depend on the Donnan properties of the adsorbent. According to Burns and Hayes (1974), an imaginary boundry can be drawn around spherical or coiled HA macromolecules encompassing a certain volume of solvent. This boundary can behave as a semipermeable membrane. Burns and Hayes (1974) suggested that in order to evaluate completely Donnan effects in ion exchange systems involving HA it would be necessary to know the volume of solution enclosed by the hypothetical membranes, that surround the polymer molecules. The approach outlined by Burns and Hayes (1974) warrants further study in its application in the organocation-HA adsorption studies. The Donnan effects will be insignificant in the presence of an excess of diffusible electrolytes in the water-polyelectrolyte system (Burns and Hayes, 1974).

(6) Ligand exchange - Adsorption by this mechanism involves replacement of one or more ligands by the adsorbent molecule. The necessary condition being that the adsorbent molecule be a stronger chelating agent than the replaced ligands. This type of mechanism may be involved for the binding of *s*-triazines on the residual transition metals of HA (Hamaker and Thompson, 1972). In ligand exchange, partially chelated transition metals may serve as possible sites for adsorption (Hayes, 1970). The pesticide molecule may displace water of hydration acting as ligand.

Coordination type of bonding may be quite important in determining the fate and behavior of pesticides in soil. Certain ligands form coordination complexes with various metals on clay minerals (Dowdy and Mortland, 1967). It was shown that urea was held on Cu^{2+}, Mn^{2+}, and N^{2+}-montmorillonite by means of coordinate coovalent bond involving carbonyl groups (Mortland, 1966). Russell et al. (1968a) demonstrated the coordination of aminotriazole to Ni^{2+} and Cu^{2+} cations on montmorillonite.

On the basis of infrared and x-ray analysis, Saltzman and Yariv (1976) demonstrated that parathion sorbed by montmorillonite coordinated through water molecules with the metallic cations in the interlayer space of the clay. Parathion became directly coordinated with the monovalent cations when the clay-parathion complexes were dehydrated. The main interaction was observed through the oxygen atoms of the nitro group and especially for complexes saturated with polyvalent cations, although interactions through the P=S group were also observed. Adsorption of 2,4-D acid on montmorillonite may also involve coordination of the acid to exchangeable metal cations through the carboxyl group via water bridging (Dieguez-Carbonell and Pascual, 1975).

Coordination through an attached metal ion (lingand exchanged) was considered to be the main process in the adsorption of linuron by clay minerals saturated with different cations (Hance, 1971). The strong band in the infrared spectra at 1278 cm^{-1}, which is indicative of C-N stretching in the thiocarbamate compounds

(Nyquist and Potts, 1961), shifts to higher frequencies upon complexing. In the case of amide, urea, and thiourea type compounds, the C-O stretching frequency is reduced when coordination occurs through this group with metal ions, and concurrently the C-N stretching frequency increases (Nakamoto, 1963). Mortland and Meggitt (1966) showed that EPTC complexes to montmorillonite by ion dipole interactions between the carbonyl of EPTC and the exchangeable metal cations on the clay. According to these workers, the decrease in C-O stretching and increase in C-N frequency was related to the electron affinity in the cation. Khan's (1973d) data also indicate coordination of the herbicide triallate to the exchangeable cations on the clay through the oxygen of the carbonyl group. The structure of triallate involves resonance between the following forms:

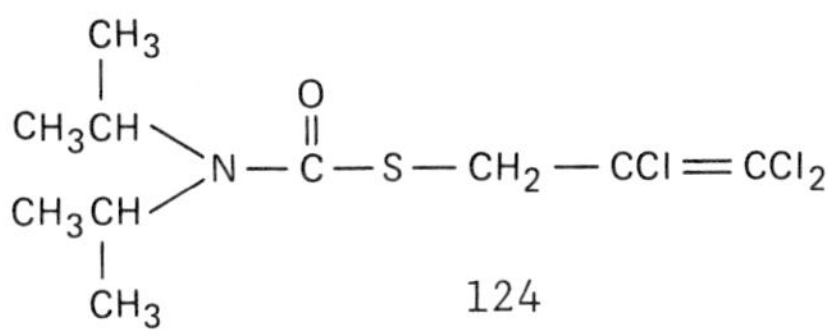

124

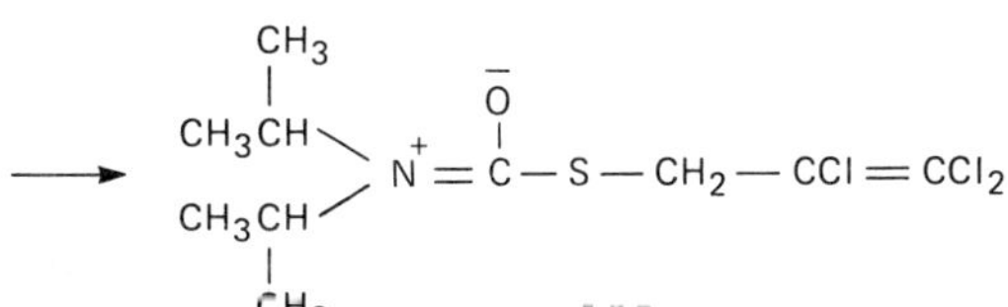

125

The contribution of structure 124 will decrease on the formation of an oxygen to metal bond. This will result in more double bond character for the C-N bond and more single bond character for C-O bond, thus increasing the C-N stretching frequency and decreasing the C-O stretching frequency.

The frequencies of C-N and C-O stretching vibrations recorded when triallate was complexed with montmorillonite saturated with various cations are shown in Table 3.4 (Khan, 1973d). In most cases, the decrease in C-O stretching frequency appears to be proportional to the electrophilic nature of the cation. Thus, the

TABLE 3.4

Vibration frequencies for triallate in the free state and when complexed with montmorillonite (Khan, 1973d)

Exchangeable cation on montmorillonite	C-O stretching (cm^{-1})	C-N stretching (cm^{-1})
Na^{+}	1595	1300
K^{+}	1598	1300
H^{+}	1590	1298
Ca^{2+}	1590	1305
Mg^{2+}	1593	1302
Cu^{2+}	1580	1302
Zn^{2+}	1587	1305
Co^{2+}	1585	1302
Ni^{2+}	1600	1300
Mn^{2+}	1590	1300
Fe^{3+}	1610	1300
Al^{3+}	1588	1300
Triallate (solid)	1665	1278

shift was greatest when the clay was saturated with Cu^{2+}, Zn^{2+} and Co^{2+}, intermediate when saturated with Ca^{2+} and Mg^{2+}, and least for Na^{+} and K^{+}. Similarly for linuron there can be two sites at which interaction with exchangeable cation is most likely to occur, the oxygen of the carbonyl group and the amide nitrogen. On the basis of infrared spectroscopy, it was shown that adsorption of linuron on montmorillonite involves coordination of the herbicide to the exchangeable cations on the clay through the oxygen of the carbonyl group (Khan, 1974d). Arnold and Farmer (1979) reported the complex formation of picloram with polyvalent cations on the exchange complex (Cu^{2+}, Fe^{3+} and Zn^{2+}) of soil. They suggested that in soils such complex reactions would most probably involve organic matter, polyvalent cations, and picloram.

3.1.5. Adsorption of Specific Types of Pesticides

Weber (1972) suggested that organic pesticides may be classified as ionic and nonionic. The ionic pesticides include cationic, basic and acidic compounds. The broad groups of pesticides classified as nonionic vary widely in their properties and include chlorinated hydrocarbons, organophosphates, substituted anilines and anilides, phenyl carbamates, phenylureas, phenylamides, thiocarbamates, acetamides, benzonitrilles and esters.

3.1.5.1. *Ionic pesticides*

(1) Cationic - This group of pesticides generally has high water solubility and ionizes in aqueous solution to form cations. The herbicides, diquat and paraquat, are the only compounds of this group that have been studied in any detail concerning the reaction with various soil constituents. In solution, they exist as divalent cations and positive charges are distributed around the molecules (Hayes et al., 1975). Diquat and paraquat are known to become inactivated in highly organic soils (Harris and Warren, 1964; O'Toole, 1966; Calderbank, 1968; Calderbank and Tomlinson, 1969; Damanakis et al., 1970; Khan et al., 1976). However, due to a slow approach to the adsorption equilibria, the inactivation process in the field has been occasionally either very slow or incomplete (Calderbank and Tomlinson, 1969). The adsorption from the solution phase by the organic matter was demonstrated by the reduction in paraquat phytotoxicity to plants grown in media containing organic soils (Scott and Weber, 1967; Coffey and Warren, 1969; Damanakis et al., 1970).

The adsorption of paraquat and diquat on soils conformed with the linear form of the Langmuir equation (Gamar and Mustafa, 1975). The adsorption maxima obtained for eight soils ranged from 17 to 47 meq/100 g.

The amount of diquat or paraquat adsorbed by soil organic matter is related to the amount of the herbicide in solution. The plot of the herbicide concentration in solution against the amount adsorbed generally has an L-shaped isotherm which levels off at a certain adsorption maximum (Calderbank, 1968; Calderbank and Tomlinson, 1969; Weber, 1972). A typical adsorption curve for paraquat on fen peat is shown in Fig. 3.9. The herbicide is completely adsorbed at low levels of application. This has often been referred to as the strong adsorption capacity region of the organic soils (Knight and Tomlinson, 1967). However, the definition of this region depends on the analytical method applied (Calderbank, 1968). Tucker et al. (1967, 1969) arbitrarily defined two types of bonding in paraquat and diquat adsorption processes by a muck soil. The 'loosely bound' paraquat is classified as

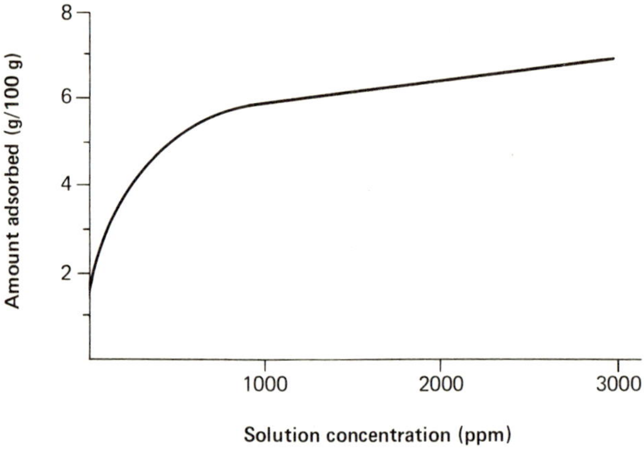

Fig. 3.9. Adsorption isotherm of paraquat on fen peat (Calderbank and Tomlinson, 1969). Published by permission of Springer-Verlag, New York.

adsorbed paraquat that can be desorbed with saturated ammonium chloride. The 'tightly bound' paraquat is classified as adsorbed paraquat that cannot be desorbed with saturated ammonium chloride, but can only be released from soil by refluxing with 18 N sulphuric acid. The 'tightly bound' capacity of muck soil for bipyridylium cations is considerably less than the 'loosely bound' capacity. Since high cation exchange capacities are characteristic of organic soils, they would have a high 'loosely bound' bipyridylium cation capacity (Tucker et al., 1967). The 'tightly bound' paraquat is not available to plants, whereas the loosely bound paraquat can potentially become available (Riley et al., 1976).

Although bipyridylium herbicides bind readily to organic materials, the binding appears to be weaker than with clay minerals. When paraquat treated organic materials were adjacent to or incorporated with clays, transfer of the herbicide to clays occurred, rendering it biologically inactive (Burns and Audus, 1970; Damanakis et al., 1970). These results demonstrate the reversibility of the binding of organo-bipyridyls complexes and the ultimate preferential adsorption by clay minerals. The higher phytotoxicity of paraquat applied to organic soils as compared to inorganic soils also indicates the relatively weak binding to organic matter (Scott and Weber, 1967; Tucker et al., 1969; Damanakis et al., 1970). Tucker et al. (1967) also suggested that bipyridyls adsorbed on soil organic fractions are loosely bound and are subject to leaching by saturated salt solution.

Adsorption of diquat and paraquat on fractionated and well characterized humic substances has been studied in greater detail (Damanakis et al., 1970; Best et al., 1972; Khan, 1973a, 1974a,b; Burns et al., 1973a,b) Khan (1973a) investigated the binding of diquat and paraquat by HA and FA by using a gel filtration technique. Paraquat was complexed by humic materials in greater amounts than was diquat, but the amount of the two herbicides complexed by HA was higher than those complexed by FA. The adsorption is influenced by the nature of the cation present on HA (Best et al., 1972; Burns et al., 1973a; Khan, 1974a). Khan (1974a) reported that the cation order for increasing adsorption for the two herbicides was nearly the same and followed the sequence: $Al^{3+}<Fe^{3+}<Cu^{2+}<Ni^{2+}<Zn^{2+}<Co^{2+}<Mn^{2+}<H^{+}<Ca^{2+}<Mg^{2+}$. The competitive ion effect between diquat and paraquat for sites on HA has been investigated by equilibrating the material with an equal molar mixture of the two herbicides (Best et al., 1972; Khan, 1974a). Table 3.5 shows the competative adsorption of paraquat and diquat on HA. The ratio of paraquat adsorbed to the total paraquat + diquat was also calculated. A value of 0.50 denotes no preference, while larger or smaller values indicate the preference in favor of paraquat or diquat, respectively. The preference was slightly in favor of paraquat. This was attributed to the relationship between surface charge density of the adsorbent and cation charge spacings, as well as steric hindrance due to cation size (Best et al., 1972).

Diquat and paraquat have been shown to be readily adsorbed by soil particles and clay minerals. Hayes et al. (1972) observed that adsorptions of paraquat and of diquat by homoionic preparations of kaolinite, illite, and montmorillonite and by Na^{+}- and Li^{+}- vermiculite preparations were complete in less than 30 minutes. Weber et al. (1965) also showed that adsorption of paraquat and of diquat by Na^{+}- kaolinite and Na^{+}- montmorillonite preparations was complete within one hour. The adsorption of bipyridylium herbicides by clays may be influenced by the lattice charge neutralizing cations. This is particularly true for vermiculite as the exchangeable cations show a marked effect on the adsorption capacity of the clay (Weed and Weber, 1968; Hayes et al., 1974). Hayes et al. (1972, 1974) investigated adsorption of paraquat and diquat by Al^{3+}-, Ca^{2+}-, Mg^{2+}-, K^{+}-, Na^{+}-, and H^{+}- saturated

TABLE 3.5

The competitive adsorption of paraquat and diquat on humic acid

Adsorbent	Herbicide added (meq/100g)		Herbicide adsorbed (meq/100g)			P[1]
	paraquat	diquat	paraquat	diquat	total	P + D
Humic Acid[2,3]	80	80	40.8	35.8	76.6	0.53
Humic Acid[2]	80	80	44.1	43.1	87.2	0.51
Humin[2]	80	80	42.1	36.1	78.2	0.54
Humic Acid[4]	50	50	39.1	39.5	78.6	0.50

[1]Ratio of paraquat (P) and diquat (D) adsorbed.
[2]Best et al. (1972).
[3]Aldrich commercial humic acid.
[4]Khan (1974a).

preparations of kaolinite, illite, montmorillonite and vermiculite. They observed that the exchangeable cations had a marked effect on the adsorption capacity of vermiculites. Adsorption reached only 80 to 90% of the cation exchange capacity (CEC) for Na^+- vermiculite, it was markedly less for some of the other cations, and it decreased in the following order: Na^+->Li^+->Sr^{2+}->Ca^{2+}->Ba^{2+}->Mg^{2+}->K^+-=NH_4^+-clay. However, the exchangeable cations had little effect on the adsorption by kaolinite, illite, and montmorillonite preparations. In all cases the herbicides were adsorbed to the CEC values of the clays and the isotherms were of the H-type (Giles et al., 1960).

In other studies, the bipyridylium cations were adsorbed up to 100% of the CEC of kaolinite and montmorillonite clays, whereas adsorption up to 90% was notified for vermiculites (Weed and Weber, 1969; Weber et al., 1965; Dixon et al., 1970). Adsorption was more complete on Na^+- saturated vermiculite than on both Ca^{2+}- and Mg^{2+} - saturated clays. X-ray diffraction studies showed that diquat and paraquat were adsorbed in the interlayer spacings of montmorillonite clay (Weed and Weber, 1968; Weber et al., 1965). Data presented by Weed and Weber (1968), and Pick (1973) on basal spacing for dried and wet complexes of paraquat- and diquat-saturated montmorillonite and vermiculite clays show that collapse of the montmorillonite lamellae occurred for the complexes of the two herbicides. Knight and Denny (1970) found that the fully saturated paraquat-montmorillonite complex could not be expanded with ethylene glycol, however, some expansion was evident for the partially saturated complex.

(2) Basic - Basic pesticides, such as *s*-triazine herbicides, readily associate with hydrogen to form a protonated species and may behave as positive counter ions. The protonated pesticide may be adsorbed via a negative site on the soil colloid (Weber et al., 1969, 1974). Evidence demonstrating the importance of soil organic matter in adsorbing *s*-triazines, in reducing their phytotoxicity, and in affecting their movement in soil has been reviewed and discussed by Hayes (1970). The adsorption of basic pesticides by soil colloid is pH dependent (McGlamery and Slife, 1966; Doherty and Warren, 1969; Weber et al., 1969). Maximum adsorption of basic pesticides, such as *s*-triazines, occurs near the pK_a of the compound. The number of protonated molecules decreases at higher pH thereby reducing the adsorption. McGlamery and Slife (1966) observed much greater adsorption of atrazine on HA under acid than under neutral conditions. In similar studies by Hayes et al. (1968), the adsorption of atrazine by hydrogen saturated muck was found to be considerably greater than that by calcium saturated muck. Gaillardon (1975) observed that terbutryn is very readily adsorbed by HA in an acid medium. Concentration of electrolytes in soil, moisture content and temperature also influence the adsorption of *s*-triazines in soil (Dao and Lavy, 1978). An extensive review concerning the adsorption of *s*-triazines by clay minerals is presented by Weber (1970a).

(3) Acidic - The acidity of this class of pesticides is mainly due to carboxylic or phenolic groups, which may ionize to produce organic anions. The activity of acidic pesticides is related to the organic matter content of soil (Upchurch and Mason, 1962; Schliebe et al., 1965; Hamaker et al., 1966; Herr et al., 1966; Scott and Weber, 1967; Grover, 1968; Keys and Friesen, 1968; O'Connor and Anderson, 1974). The magnitude of adsorption of acidic pesticides by soil colloids is much lower than that of cationic or basic pesticides (Weber, 1972). The adsorbed pesticides can readily be released to water (Harris and Warren, 1964; Weber et al., 1968). Adsorption of acidic pesticides depends on the pH of the system. At low pH levels, most of the weakly acidic herbicides are present in the molecular rather than the

anionic form. Thus, they would be adsorbed to a greater extent than stronger acid herbicides. Picloram adsorption has been shown to be poorly correlated with soil clay content but significantly correlated with soil organic matter content (Grover, 1971; Hamaker et al., 1966; Herr et al., 1966). Picloram is preferentially adsorbed in the molecular form, i.e. picloram adsorption is increased with decreasing pH (Hamaker et al., 1966; Grover, 1971). Arnold and Farmer (1979) showed that adsorption of picloram was adequately described by the Freundlich adsorption isotherms. They observed that picloram was adsorbed on soils to a much greater extent at low pH values. Thus, the increased adsorption below the pK_a of picloram (3.6) indicates a preferential adsorption of the unionized or molecular form of the herbicide. For the soil saturated with metallic cations, the order of decreasing picloram adsorption capacity was $Fe^{3+}=Cu^{2+}>Al^{3+}>Zn^{2+}>Ca^{2+}$. Picloram has been shown to be adsorbed on HA and humin largely in the form of uncharged molecules (Nearpass, 1976). Some phenolic pesticides exist as the free acid in acidic soils and may be adsorbed on organic matter. Su and Lin (1971) observed that the efficacy of PCP was strongly influenced by organic matter. PCP efficacy decreased with increase in organic matter. Positive correlations have been observed between PCP efficacy and organic matter content of soil (Tsunoda, 1965; Choi and Aomine, 1972). Choi and Aomine (1974) suggested that organic matter plays an important role in adsorption of PCP in soil. They observed that a decrease in organic matter content resulted in a decrease in adsorption of PCP.

The acidic pesticides will not be adsorbed by either montmorillonite or kaolinite at high pH. However, the adsorption could be slightly positive at low pH. According to Bailey et al. (1968), the adsorption of some acidic herbicides appears to be more closely related to the pH of the bulk solution. Adsorption of the molecular species alone occurs at suspension pH values 1 to 2 units below the pK_a of the compound.

(4) Miscellaneous ionic pesticides. Some of the ionic pesticides do not fall into the above described categories of compounds. Included in this group are bromacil, terbacil, isocil, oryzalin, DSMA, and cacodylic acid. They exhibit weak acidic or basic properties and may also possess certain functional groups in the molecule. The latter cause them to behave differently from cationic, basic or acid pesticides. The uracil herbicides are partially

adsorbed by soil organic matter (Burnside et al., 1969; Rhodes et al., 1970). It has been shown that DSMA is readily adsorbed by various clay minerals and soil particles (Dickens and Hiltbold, 1967).

3.1.5.2. *Nonionic pesticides*

Pesticides included in this category vary widely in their properties and do not ionize significantly in aqueous or soil systems. Adsorption of nonionic pesticides on soil colloids depends mainly upon the chemical properties of the compounds and the types of soil surface involved. In the following paragraphs the adsorption of the broad group of pesticides classified as nonionic on soil colloid is discussed.

(1) Chlorinated hydrocarbons - The effect of soil organic matter on the insecticidal activity of several chlorinated hydrocarbons was first observed by Fleming (1950), Fleming and Maines (1953, 1954), and Edwards et al. (1957). Later investigations confirmed the influence of soil organic matter on the bioactivity of both volatile and nonvolatile chlorinated hydrocarbons, (Bowman et al., 1965; Weil et al., 1973).

Many investigators found that the retention and inactivation of DDT in soil was related to the organic matter content of the soil. Shin et al. (1970) observed that DDT adsorption in soil was greater in more humified soil organic matter. Pierce et al. (1974) investigated DDT adsorption to a marine sediment, sediment fractions, clay and HA suspended in sea water. The humic fraction was found to have a greater adsorbing capacity than the clay or sediment. Removal of humic fractions from sediment reduced the adsorption capacity to less than 50% of the original sediment sample. Pierce et al. (1974) concluded that suspended humic particulates may be important agents for transporting chlorinated hydrocarbons through the water column and for concentrating them in sediments. Movement of DDT in forest soils has been attributed to its association with HA and FA fractions of soil organic matter (Warshaw et al., 1969; Ballard, 1971). Warshaw et al. (1969) observed that DDT was more soluble in sodium humate than in distilled water. The increased solubility of the insecticide was related to the effect of humate on lowering the surface tension of water.

The lipid fraction of soil organic matter has also been implicated in the adsorption of DDT (Pierce et al., 1971). It has been suggested that the adsorption of nonpolar pesticides on soil organic matter is mainly due to pesticide-lipid interaction.

(2) Organophosphates - Adsorption of organophosphate pesticides is related to the organic matter and clay content of soils (Kirk and Wilson, 1960; Swobada and Thomas, 1968; Felsot and Dahm, 1979). The bioactivity of phorate was found to decrease with an increase in organic matter content of soils (Kirk and Wilson, 1960). Soil moisture affects the adsorption of organophosphates and chlorinated hydrocarbons in a similar fashion. Saltzman et al. (1972) observed that in aqueous solution parathion had a greater affinity for organic than for mineral adsorptive surfaces in soils. Parathion adsorption by soils can be described by the Freundlich empirical equation and the adsorption is not totally reversible (Yaron and Saltzman, 1978; Wahid and Sethunathan, 1978). Since parathion retention by organic colloids is stronger than by mineral surfaces, the organic matter is the main factor affecting parathion release from the sorbed state to the soil solution (Yaron and Saltzman, 1978). Inorganic soil constituents influence parathion sorption in soils with <2% organic matter, but their role is apparently masked by organic matter at levels above 2% (Wahid and Sethunathan, 1978). Humic materials, such as FA can increase or decrease certain organophosphorous insecticides adsorption by montmorillonite clay suspensions, depending on the humic material concentration and the saturating cations (Bowman, 1978). Adsorption of organophosphorous insecticides on clay minerals and soils is also influenced by the saturating cations (Chopra et al., 1970; Bowman, 1973; Harris and Bowman, 1976; Bowman and Sans, 1977; Yaron, 1978).

The hydration status of clay minerals affect their adsorption capacity for organophosphorus compounds. Fig. 3.10 demonstrates the decrease in the amount of parathion adsorbed by an attapulgite from hexane solution as affected by the hydration water of the mineral (Yaron, 1978). The clay was equilibrated previously up to a relative humidity of 98 percent. The high adsorption in a dry system is attributed to the effective competition of polar parathion molecules with nonpolar hexane molecules for the adsorption site. In partially hydrated systems, parathion molecules are unable to replace the strongly adsorbed water molecules, so that parathion adsorption occurs on water free surfaces only. This

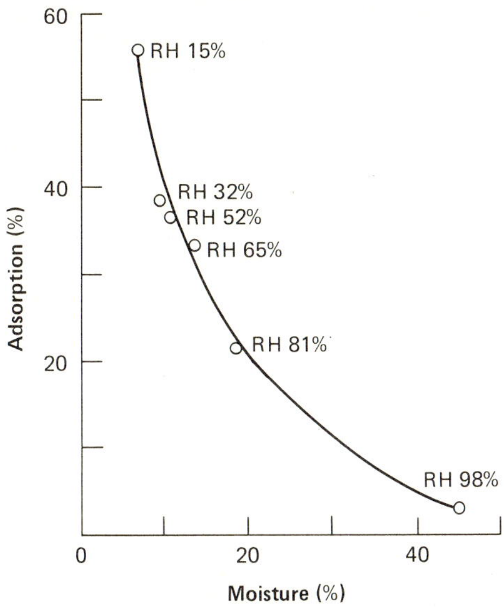

Fig. 3.10. Adsorption of parathion by attapulgite from hexane solution as affected by initial hydration status of the mineral (Yaron, 1978).

results in an apparent decrease in the adsorption capacity of attapulgite for parathion. However, it is possible that the apparent decrease may be due to time required by the parathion molecule to diffuse through water to the active adsorption site. Thus, by increasing sufficiently the time of contact between the adsorbent and adsorbate, a similar adsorption capacity may be reached for dry and hydrated parathion-hexane-attapulgite systems (Yaron, 1968). Bowman et al. (1970) observed that malathion was adsorbed as a double layer in the interlayer spacing of montmorillonite clay. The possible presence of parathion in the interlayer space of sodium montmorillonite has been recently demonstrated by x-ray analysis (Biggar et al., 1978). Getzin and Chapman (1959) observed no significant adsorption of phorate on kaolinite.

Various organic matter fractions were found to adsorb parathion (Leenheer and Ahlrichs, 1971). Furthermore, it was observed that organic matter with H^+ on exchange sites adsorbed significantly larger amounts of the insecticide than with Ca^{2+} on the exchange sites. In a recent study, Khan (1977a) investigated adsorption of fonofos on HA saturated with different cations. The amount of the insecticide adsorbed was affected by the cation with which the HA was saturated. This suggests that the mobility and persistence of fonofos in soils will be partly a function of adsorption on humic materials. Grice et al. (1973) showed that HA has a high affinity for organophosphorus compounds. Their experiments gave an adsorption capacity of about 30g of dimefox per 100g HA.

(3) Substituted anilines - The substituted anilines are readily adsorbed by soil organic matter (Lambert, 1967; Hollist and Foy, 1971; Weber et al., 1974). Harvey (1974) measured the extent and strength of adsorption of 12 substituted aniline herbicides by a silt loam soil and extrapolated results to estimate equilibrium concentrations at field moisture capacity. Jacques and Harvey (1979) observed that adsorption of benefin, dinitramine, fluchloralin, oryzalin, profluralin and trifluralin on 10 Wisconsin soils followed the Freundlich isotherms and the adsorption was related more closely to soil organic matter than to the other soil chemical and physical properties. The phytotoxicity of benefin was found to be significantly correlated with the organic matter content of soil (Weber et al., 1974). According to Lambert (1967), the adsorption of some substituted anilines by organic matter is related to the parachor of the compounds; larger molecules are adsorbed more than smaller molecules. The herbicide trifluralin was adsorbed in small amounts by montmorillonite and kaolinite clays (Coffey and Warren, 1969).

(4) Phenylureas - The herbicidal activity of phenylureas is related to the organic matter content of the soils (Upchurch and Mason, 1962; Savage and Wauchope, 1974; Weber et al., 1974; Carringer et al., 1975; Chang and Stritzke, 1977). The adsorption of linuron by organic soils is increased with decomposition (Morita, 1976). The pH of the system did not affect adsorption of phenylureas significantly (Yuen and Hilton, 1962; Hance, 1969a). Hance (1965a) observed a competition between water and diuron for adsorption sites, and also that diuron was a more effective competitor at soil organic matter surfaces than at soil mineral matter surfaces.

The adsorption of linuron by organic matter and clay minerals is affected by the cation with which the adsorbent is saturated. Thus, the adsorption of linuron by (1) peat (Hance, 1971), (2) HA (Khan and Mazurkewich, 1974), (3) bentonite (Hance, 1971) and (4) montmorillonite (Khan, 1974d) saturated with various cations decreased in the following order:

(1) $Ce^{4+} > Fe^{3+} > Cu^{2+} > Ni^{2+} > Ca^{2+}$
(2) $H^{+} > Fe^{3+} > Al^{3+} > Cu^{2+} > Ca^{2+} > Zn^{2} > Ni^{2+}$
(3) $Fe^{3+} > Ce^{4+} > Cu^{2+} > Ni^{2+} > Ca^{2+}$
(4) $Al^{3+} \geq Cu^{2+} \geq Ni^{2+} > H^{+} > Mg^{2+}$

Phenylurea derived chloroaniline residues in soil were found to be immobilized by adsorption on humic materials (Hsu and Bartha, 1974a, b; Bartha and Hsu, 1976). Chemical attachment of chloroanilines to humic substances occurs both in a hydrolyzable and in a nonhydrolyzable manner (Hsu and Bartha, 1974a).

Relatively low adsorption of monuron and diuron from aqueous solutions by montmorillonite, illite and kaolinite clay minerals has been observed (Frissel and Bolt, 1962). Adsorption of several phenylureas on clay minerals has also been reported by other workers (Geissbuhler et al., 1963; Harris and Warren, 1964; Bailey et al., 1968). It was shown that adsorption of phenylureas by clay minerals was slightly greater under acid conditions than under basic or neutral conditions (Frissel and Bolt, 1962; Harris and Warren, 1964). Furthermore, the adsorption was much greater on hydrogen saturated montmorillonite than on sodium saturated montmorillonite (Bailey et al., 1968).

(5) Phenylcarbamates and carbanilates - Chlorpropham and propham inactivation is related to the organic matter content of the soil (Upchurch and Mason, 1962; Harris and Sheets, 1965). Chlorpropham is adsorbed reversibly by muck (Harris and Warren, 1964; Hance, 1967) and its phytotoxicity reduced by the organic matter added to the soil (Scott and Weber, 1967). Carbaryl, an insecticide, was shown to be adsorbed by various organic matter fractions (Leenheer and Ahlrichs, 1971). Chlorpropham and propham are adsorbed by montmorillonite clay (Harris and Warren, 1964; Schwartz, 1967; Coffey and Warren, 1969). However, the amounts adsorbed on kaolinite and illite clays are insignificant (Schwartz, 1967).

(6) Substituted anilides - The adsorption of substituted anilides on soil colloids has not been studied in detail. Recently, butralin and profluralin were shown to be strongly adsorbed by soil organic matter (Carringer et al., 1975). Bailey et al. (1968) observed that dicryl, solan and propanil were adsorbed in small amounts by Na-montmorillonite but to a greater extent by

H-montmorillonite. The water solubilities of the compounds were not related to the amounts adsorbed.

(7) Phenylamides - In leaching experiments, it was observed that diphenamide moved less as the organic matter content of the soil was increased (Deli and Warren, 1971). It was reported that up to 90% of the 3,4-dichloroaniline released during the biodegradation of several phenylamide herbicides becomes unextractable by solvents due to binding to the soil organic matter (Hsu and Bartha, 1976). Diphenamide was found to be adsorbed in moderate amounts by muck and charcoal (Coffey and Warren, 1969).

(8) Thiocarbamates, carbothioates, and acetamides - Movement of certain thiocarbamates is considerably less in soil as the organic matter content increases (Gray and Weierich, 1968; Koren et al., 1968; 1969). Increase in organic matter content results in increased adsorption of thiocarbamates and acetamides (Ashton and Sheets, 1959; Deming, 1963; Koren et al., 1968, 1969; Carringer et al., 1975). Organic matter content of soil is related to the herbicidal activities of thiocarbamate and acetamide (Ashton and Sheets, 1959; Jordan and Day, 1962). Thiocarbamates and acetamides are readily adsorbed by certain clay minerals (Mortland and Meggitt, 1966; Koren et al., 1969). The adsorption isotherms of the insecticidal carbamate, aldicarb, for three soils and their organo-clay constituents isolated from these soils indicated that both negative and positive adsorption occurred in these systems (Supak et al., 1978).

(9) Benzonitriles - The benzonitrile herbicide, dichlobenil was adsorbed on soil organic matter (Massini, 1961). Lignin also was reported to adsorb dichlobenil from aqueous solution (Briggs and Dawson, 1970).

3.1.6. Adsorption of Pesticides by Organo-Clay Complexes

The presence of organic matter-clay complexes in most of the mineral soils needs to be considered in evaluating the importance of organic matter in pesticide adsorption. Stevenson (1976) quoted Walker and Crawford (1968) indicating that up to an organic matter content of about 6% both mineral and organic surfaces are involved in adsorption. However, at higher organic matter contents, adsorption will occur mostly on organic surfaces. Stevenson (1976) pointed out that the amount of organic matter required to coat the clay will depend on the soil type and the kind and amount of clay that is present.

The intimate association of organic matter and clay may cause some modification of their adsorptive properties, or they may complement one another in the role of pesticide adsorption (Pierce et al., 1971; Niemann and Mass, 1972). Only recently have attempts been made to study the adsorption of pesticides by organic matter-clay complexes. Burns (1972) pointed out that a humus-clay micro-environment is a site of high biological and nonbiological activity and it is here that we need to look for the basic information concerning soil-pesticide interactions.

The adsorptive capacities of sedimentary organomineral complexes for lindane and parathion were found to be much greater than these of the corresponding mineral fraction (Graetz et al., 1970). Furthermore, the extent of adsorption was related to the organic carbon content of the complex. Wang (1968) obtained similar results for the adsorption of parathion and DDT on organo-clay fractions. Miller and Faust (1972) investigated the adsorption of 2,4-D by several organo-clay complexes. The latter were prepared by treating dimethylbenzyl octadecylammonium chloride and various benzyl and aliphatic amines with Wyoming bentonite. It should be noted, however, that the nature of the organic matter in soil differs profoundly from the organic compounds used by Miller and Faust (1972). Thus, the adsorption behavior of their organo-clay complexes may differ significantly from those found in soil. Khan (1974c) investigated the adsorption of 2,4-D by a FA-clay complex prepared by treating FA with Na-montmorillonite. This FA-clay complex was similar to the naturally occurring organo-clay complexes found in soil (Kodama and Schnitzer, 1971). Khan (1974c) observed that the FA-clay complex adsorbed about 6.5 and 5.2 μmole of 2,4-D per g of complex at 5° and 25°C, respectively.

Hance (1969a) suggested that in soil, clay and organic matter associate in such a manner that little of the clay mineral surface will be accessible to pesticide molecules. Thus, the contribution to adsorption of the clay fraction in soils would be much less than studies with the isolated mineral would indicate. On the other hand, Mortland (1968) is of the opinion that organic compounds in soil organic matter, upon interaction with clay, may facilitate and stabilize adsorption of pesticides beyond that observed in purely inorganic clay systems. In order to shed some light on these rather contradictory speculations, Khan (1973c) estimated the amounts of diquat and paraquat adsorbed by montmorillonite and an organo-clay complex when increasing amounts of the herbicide was added to each system (Table 3.6). The organo-clay complex constituted 62% and 38% of montmorillonite and FA, respectively. It was observed that diquat and paraquat were adsorbed in considerably greater amounts by the clay when present in the form of organo-clay complex. Thus, when 1200 μmole of the herbicide was initially added to the organo-clay complex, 1 g montmorillonite adsorbed 532 and 597 μmole of diquat and paraquat, respectively. The corresponding values for diquat and paraquat adsorption by 1 g montmorillonite in pure clay system were 420 and 445 μmole, respectively (Table 3.6). It appears that FA, which is the most prominent humic compound in soil solution on interacting with clay minerals will facilitate the adsorption of pesticides on clays in soils.

TABLE 3.6

Adsorption of diquat and paraquat (μmole/g) by montmorillonite and an organo-clay complex (Khan, 1973c)

Pesticide added μmole	Amount adsorbed by montmorillonite		Amount adsorbed by organo-clay complex[1]	
	Diquat	Paraquat	Diquat	Paraquat
200	200	200	200	200
400	400	400	305	310
600	410	430	320	340
800	410	440	330	360
1000	420	445	330	370
1200	420	445	330	370

3.2 MOVEMENT IN SOIL

Movement of a pesticide in the soil environment may occur while in solution or adsorbed on migrating particulate matter, or by volatilization. Movement through soil in the solution phase may involve the diffusion and mass flow processes. The relative importance of diffusion processes in soil water and air depends in part on the solubility and vapor pressure of a pesticide. Diffusion is the process by which matter is transported as a result of random molecular motions caused by their thermal energy. Thus, there is a net movement from positions of higher concentrations to positions of lower concentrations. Mass flow occurs as a result of external forces acting on the carrier for the pesticide in question. Leaching of pesticides is usually considered synonomous with mass flow, although diffusion occurs simultaneously. The summation of diffusion and mass flow processes determines the total rate of movement of a pesticide in soil.

This section begins with a description of the two general processes, diffusion and mass transfer, and is followed by a discussion on volatilization and run off.

3.2.1. Diffusion

Diffusion influences the distribution pattern of pesticides in soil. According to Fick's laws of diffusion:

$$J = -D \frac{\partial C}{\partial x} \qquad (3.11)$$

where J is the quantity of transfer per unit cross sectional area per unit time, D is the diffusion coefficient, C is the concentration, and x the space coordinate measured normal to the section. For a simple system such as diffusion through water, the Fick's law equation can be represented by:

$$\frac{C}{t} = D \frac{\partial^2 C}{\partial x^2} \qquad (3.12)$$

The approach of Shearer et al. (1973) to diffusion analysis in the soil system was to incorporate soil variables such as bulk density and water content into eqs. 3.11 and 3.12 so that the diffusion coefficient measured can be extrapolated to other soil conditions. In their mathematical development for diffusion assumption was made that the pesticide was volatile, so that diffusion occurs both in the vapor and non vapor phases. Furthermore, diffusion in the non vapor phase was assumed to occur in solution and at the solution-solid and solution-air interface. The relationship developed by Shearer et al. (1973) helps in making qualitative assessments of diffusion.

The reader is referred to Hamaker (1972) and Shearer et al. (1973) for a detailed treatment of diffusion of organic pesticides in soil.

It is well known that diffusion of pesticides can occur both in the vapor and in the non vapor phases. The latter can occur in solution or at the air-water or air-solid interface. Distribution of some pesticides into smaller pores, aggregates, and blocked pores of soil is dependent on diffusion. Volatile fumigants diffuse rapidly through a porous media except when the water content is high.

Relatively few studies of pesticide movement have dealt directly with diffusion. In general, the diffusion coefficients (D) of pesticides are 1 to 3 x 10^4 times greater in air than in water. Thus, pesticides with a water/air ratio under 1 x 10^4 should diffuse primarily through air, whereas those with ratios over 3 x 10^4 should diffuse principally through water (Goring, 1967). A number of soil and environmental factors influence the diffusion of pesticides in soil. These factors are diffusion coeffient, solubility, vapor density, adsorption, bulk density, soil water content and porosity. Graham-Bryce (1969) derived an equation showing how soil factors affect pesticide diffusion:

$$D = \frac{D_L V_L f_L}{b \quad + V_L} \tag{3.13}$$

where D_L is the diffusion coefficient in the free solution, V_L is the fraction of soil occupied by the liquid phase, f_L is the tortuosity factor for a soil, b is the slope of adsorption isotherm, and is the bulk density. Some of the parameters which influence the diffusion of pesticides in soils are discussed below.

3.2.1.1. Adsorption

According to eq. 3.13, increased adsorption should reduce diffusion. Such a relationship has been found by Walker and Crawford (1970) for propazine and prometryn. Lindstrom et al. (1968) provided evidence that the effective diffusion coefficient of 2,4-D in a number of soils was reduced by adsorption of the herbicide by soil. For three *s*-triazines Lavy (1970) showed that factors normally correlated with increased adsorption, such as organic matter, have been correlated with increased diffusion. The diffusion of dieldrin in relatively dry soils was found to increase from 3.8 mm^2/week in a fine sandy loam to 9.8 mm^2/week in a clay soil (Farmer and Jensen, 1970).

3.2.1.2. *Soil water content*

Shearer et al. (1973) investigated the diffusion of lindane through Gila silt loam soil and measured the vapour and non vapor diffusion components as a function of soil water content. They observed that essentially no diffusion occurred in dry soil, but increased rapidly with increasing water content reaching to a maximum at about 4% water content. With further increase in water content, a decline in total diffusion was observed until at 30% water content when an increase in diffusion occurred with increasing water. A slight decrease in vapor diffusion was observed as water content increased from 4 to 20% and then decreased rapidly at water contents above 20%. Ehlers et al. (1969) determined the ratio of diffusion occurring in the vapor and non vapor phases at two water contents in Gila silt loam. Approximately half of the lindane diffused in the vapor phase at 10% soil water content, whereas at near saturation diffusion was totally in non vapor phase. Graham-Bryce (1969) observed a rather rapid increase in the diffusion coefficient of dimethoate as the soil water increased. However, the disulfoton diffusion coefficient remained relatively constant over the entire soil water content range used. The apparent diffusion coefficient of many herbicides tends to increase with an increase in soil water content (Lavy, 1970; Scott and Phillips, 1972). The effect of water content on diffusion of pesticides under dry conditions has also been reported in the literature (Barlow and Hadaway, 1955, 1958; Farmer and Jensen, 1970). Diffusion of dinitroaniline herbicides is affected by soil water (Jacques and Harvey, 1979). Diffusion of trifluralin, profluralin and benefin decreased as soil water increased. Diffusion of dinitramine and fluchloralin did not change significantly with change in water content, while diffusion of oryzalin increased at the highest soil water content.

3.2.1.3. *Temperature*

Diffusion coefficient and vapor density tend to increase with temperature. The overall effect of increasing temperature is an increase in diffusion. Lavy (1970) observed a decrease in the diffusion coefficient for atrazine, propazine, and simizine when the temperature in several soils was decreased from 25° to $5^{o}C$. Call (1957b) reported a decrease in the diffusion coefficient of EDB with decrease in temperature. Ehlers et al. (1969) observed an exponential increase in the apparent diffusion coefficient for lindane with increase in temperature.

3.2.1.4 *Bulk density*

Increase in soil bulk density results in a decrease of the diffusion coefficient. Farmer et al. (1973) reported that volatilization of dieldrin tended to decrease as bulk density increased. They observed that the principal effect of bulk density was that of limiting the vapor phase movement of dieldrin to the soil surface. A decrease in the apparent diffusion coefficient of lindane from 16.5 to 7.5 mm^2/week was observed when the bulk density of the silt loam soil was increased from 1.00 to 1.55 g/cm^3 (Ehlers et al., 1969). Call (1957a) observed a decrease in the measured apparent diffusion coefficient of EDB due to an increase in the bulk density of a loamy sand soil.

Diffusion in a silt loam soil was adequately described by eq. 3.12 for the herbicide trifluralin (Bode et al., 1973a,b). Diffusion coefficient D was constant regardless of concentration or time. For bulk densities between 1.2 and 1.4 g/cm^3, the magnitudes of vapor and solution diffusion were similar, below 1.2 g/cm^3, vapor diffusion was more important. Vapor diffusion was decreased about 50% for each 10% decrease in air filled porosity (Bode et al., 1973a).

Bode et al. (1973b) reported diffusion of trifluralin as a function of soil moisture, soil temperature, and bulk density (Fig. 3.11). The maximum diffusion in a compact soil occurred at about 10% soil moisture content (Fig. 3.11a). At low bulk density maximum diffusion was shifted to higher moisture contents and was approximately 2.5 times higher than values from soil at high bulk density (Fig. 3.11b). Diffusion greatly increased with temperature

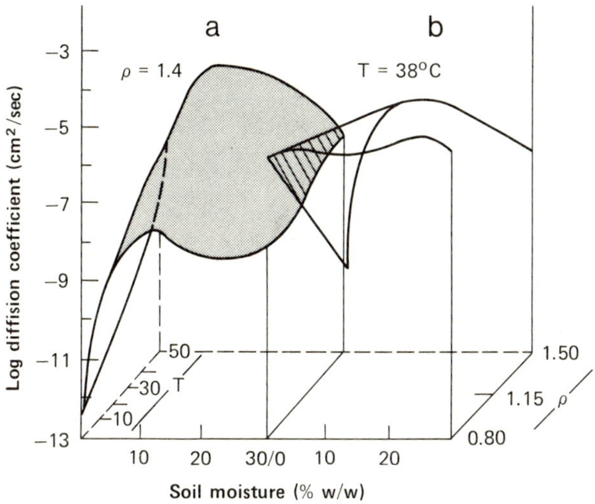

Fig. 3.11. Response surfaces for trifluralin diffusion coefficients in Mexico silt loam using a 15-term prediction model with a constant (a) bulk density, ρ, of 1.4 g/cm^3, or (b) soil temperature, T, of 38°C (Bode et al., 1973b). Published by permission of the American Society of Agronomy, Crop Science Society of America, and the Soil Science Society of America.

(Fig. 3.11a). When the air filled fraction of the soil void volume was reduced below 40% vol/vol by either compression or addition of moisture, diffusion of trifluralin began to decrease.

3.2.2. Mass Flow

Mass flow occurs as a result of external forces on water, air, or soil particles that serve as a carrier for the pesticide. Therefore, knowledge of factors affecting water, air, and soil movement is essential in order to understand the mass flow of pesticides in soil. Furthermore, it is also important to understand factors that affect the addition or removal of pesticides from these carriers.

3.2.2.1. Water as a carrier

The pesticide may be associated with water as a solution, suspension, or emulsion. Mass flow by water through a soil profile will depend on the direction and rate of water flow and the sorption characteristics of the pesticide with soil. The latter controls the distance of movement and the maximum pesticide concentration. Various models have been proposed to predict the mass transport of pesticides through the soil (Oddson et al., 1970; Letey and Oddson, 1972; Leistra, 1973; Letey and Farmer, 1974). Helling (1970) summarized the field and laboratory techniques used in predicting the distribution of pesticides through a soil profile. The field methods involve residue analysis with depth and lysimeter experiments (Riekerk and Gessel, 1968). The laboratory methods include soil columns containing soil and applied pesticide (Geissbühler et al., 1963; Hilton and Yuen, 1966; Davidson et al., 1968; Davidson and Chang, 1972; Huggenberger et al., 1972; Hornsby and Davidson, 1973, van Genuchten et al., 1974). Data obtained from column studies have been used to estimate the movement of pesticides in the field (Swoboda and Thomas, 1968). However, factors such as variations in profile characteristics, flow rate, amount of soil water, surface evaporation, etc. may restrict the comparison.

Adsorption appears to be the most important factor influencing the mass transport of a pesticide through soil by water. Several workers have observed an inverse relationship between adsorption and movement of pesticides by water through soil (Ashtan, 1961; Hamaker et al., 1966; Harris, 1966, 1967a,b, 1969; Guenzi and Beard, 1967). The nature of pesticides is also important inasmuch as they affect adsorption.

Thin layer chromatographic (TLC) techniques have also been used to measure pesticide mobility through soils (Helling and Turner, 1968; Helling, 1971a, b,c; Singhal and Bansal, 1978). However, absolute movement on TLC plates cannot be transposed directly to field or soil column experiments.

3.2.2.2. *Soil as a carrier*

Pesticides may become intimately associated with soil particles by adsorption. The soil particles may act as a carrier when moved by water or air which is referred to as water or wind erosion. Pesticides most likely to be removed by erosion are those that are not mobile. The amount of pesticide moved by erosion will depend upon the amount adsorbed by the transporting soil.

Various workers have studied the movement of pesticides caused by water erosion or runoff. Barnett et al. (1967) observed that formulation of 2,4-D as the amine salt greatly reduced runoff when compared with esters of the herbicide. Trichell et al. (1968) studied the loss of 2,4,5-T, dicamba and picloram from sodded and plowed plots. In the sodded plots, applied herbicides were moved in the initial runoff. Four months later, however, losses were reduced to $<1\%$ of the initial value. The concentration of herbicides in water during the first 24 hours was sufficient to cause some damage in a bioassay experiment. White et al. (1967) studied the loss of atrazine applied to fallow soil under simulated rainfall conditions. Most of the atrazine was lost during the early part of runoff and less at the later stage of runoff. Epstein and Grant (1968) investigated the removal of chlorinated insecticides from field plots in runoff from natural rainfall. Higher amounts of DDT, endrin and endosulfan were removed when the rain came very shortly after application. DDT was more persistent and appeared in higher concentration in the runoff than the other two insecticides investigated. Hindin et al. (1966) measured runoff of DDT, ethion, and diazinon insecticides from a coarse silt loam soil. Less than 0.01% of pesticide applied was recovered in runoff water plus silt.

Wind can move soil particles to great distances. Thus, the adsorbed pesticides can be transported over large distances by this mechanism. The wind erosion for several herbicides was demonstrated by Menges (1964). According to Cohen and Pinkerton (1966), pesticides were found in rain water in the range of 0.02-1.18 ppb of organic chlorine. They speculated that the pesticides were associated with dust particles in the air.

Helling et al. (1971) ranked the relative mobility of a number of pesticides in soils. The data shown in Table 3.7 are based on many references. Compounds of class I are immobile while those of Class V are very mobile. Within each class, pesticides are ranked in estimated decreasing order of mobility. Helling et al. (1971) suggested that pesticides are generally of intermediate to low mobility, although acidic compounds are relatively mobile. Phenylureas and *s*-trazines belong to the mobility class II or III, and chlorinated hydrocarbon insecticides are usually least mobile, preceded somewhat by organophosphate insecticides.

TABLE 3.7

Relative mobility of pesticides in soils (Helling et al., 1971)

Mobility class				
I	II	III	IV	V
Neburon	Siduron	Propachlor	Picloram	TCA
Chloroxuron	Bensulide	Fenuron	Fenac	Dalapon
DCPA	Prometryn	Prometone	Pyrichlor	2,3,6-TBA
Lindane	Terbutryn	Naptalam	MCPA	Tricamba
Phorate	Propanil	2,4,5-T	Amitrole	Dicamba
Parathion	Diuron	Terbacil	2,4-D	Chloramben
Disulfoton	Linuron	Propham	Dinoseb	
Diquat	Pyrazon	Fluometuron	Bromacil	
Chlorphen-amidine	Molinate	Norea		
Dichlormate	EPTC	Diphenamid		
Ethion	Chlorthiamid	Thionazin		
Zineb	Dichlobenil	Endothall		
Nitralin	Vernolate	Monuron		
TH-1568A	Pebulate	Atratone		
Morestan	Chlorpropham	Atrazine		
Isodrin	Azinphos-methyl	Simazine		
Benomyl	Diazinon	Ipazine		
Dieldrin		Alachlor		
Chloroneb		Ametryne		
Paraquat		Propazine		
Trifluralin		Trietazine		
Benefin				
Heptachlor				
Endrin				
Aldrin				
Chlordane				
Toxaphene				
DDT				

3.3. VOLATILIZATION

Certain pesticides may be distributed throughout the soil pro-ile by vapor phase movement and eventually lost via surface evaporation. Volatilization loss rate of pesticides in soil is related to the vapor pressure of the pesticide within soil and its rate of movement to the evaporating surface. The characteristic saturation vapor pressure of every pesticide varies with temperature. The vapor pressure is also influenced by adsorption on soil (Spencer et al., 1969; Spencer and Cliath, 1969, 1970a,b, 1972). Spencer (1970) pointed out that the magnitude of the adsorption effect, or reduction of the vapor pressure, of a pesticide in soil is dependent mainly upon the nature of the pesticide, its concentration in soil, soil water content, and soil properties such as organic matter, clay content and pH.

An inverse relationship between the rate of pesticide volatilization, and soil organic matter content has been reported by several workers (Harris and Lichtenstein, 1961; Guenzi and Beard, 1970). Fang et al. (1961) observed that EPTC loss was greater from soils low in organic matter. Kearney et al. (1964) observed that among several *s*-trazines, the largest vapor losses were obtained from prometone and appeared to be inversely related to the amount of clay and organic matter. Spencer (1970) observed an inverse relationship between vapor density and organic matter content regardless of the fact that the clay content was inversely related to the organic matter content in most of the soils (Table 3.8). It appears that clay plays only a minor role in the adsorption of such weakly polar compounds when sufficient water is present to cover the mineral surface. With the drier soils, dieldrin vapor density was greatly decreased, but the inverse relationship between organic matter and vapor density was still apparent.

TABLE 3.8

Effect of organic matter and clay content on vapor density of dieldrin (10 ppm) at 30°C in wet and dry soils (Spencer, 1970)

Soil texture	Organic matter %	Clay %	Vapor density			
			Wet[1]		Dry[2]	
			ng/l	d/d_o[3]	ng/l	d/d_o[3]
Fine sandy loam	0.19	16.3	175	0.87	1.7	0.008
Clay	0.20	67.3	200	1.00	2.9	0.014
Loam	0.58	18.4	52	0.26	0.7	0.004
Sandy loam	1.62	10.0	32	0.16	0.4	0.002
Clay loam	2.41	33.4	32	0.16	0.6	0.003

[1]Wet - approximately 2 atm. matrix suction
[2]Dry - in equilibrium with 50% relative humidity
[3]Relative vapor density

The concentration of a pesticide in soil is related to its vapor density. The vapor densities of dieldrin and lindane in a silt loam soil wet to 10% water content increased with an increase in pesticide concentration to a level where the soil air was saturated (Spencer et al., 1969; Spencer and Cliath, 1970b). It was demonstrated that vapor densities of *p,p'*-DDT, *o,p'*-DDT, *p,p'*-DDE, and *o,p'*-DDE are the function of concentration in a silt loam soil (Spencer and Cliath, 1972). Thus, the pesticide concentration in soil influences the vapor density, which in turn controls the vapor loss.

Water plays an important role in the volatilization of pesticides from soil. Pesticides volatilize much more rapidly from wet than from dry soil (Fang et al., 1961; Harris and Lichtenstein, 1961; Deming, 1963; Kearney et al., 1964; Bowman et al., 1965; Gray and Weierich, 1965; Parochetti and Warren, 1966; Guenzi and Beard, 1970; Willis et al., 1971). Evaporation of water could enhance pesticide volatilization by the 'wick' evaporation (Hartley, 1969). Thus, as the water evaporates from the surface, the water-pesticide solution moves towards the evaporating surface by capillary action, thereby enhancing pesticide loss by water evaporation. However, in a recent study, Saltzman and Kliger (1979) observed smaller losses of the fumigant DBCP from wet than from dry soil. They attributed this reduced volatilization loss to adsorption, especially in soils with a high clay content, and the possibility of water acting as a soil cover when added after DBCP application. Spencer et al. (1973) reported that the volatalization rate of soil incorporated lindane and dieldrin was controlled by diffusion of the pesticide and by mass flow of water to the soil surface. A simplified relationship based on diffusion can be used to calculate the volatilization losses:

$$Q_t = 2C_o(Dt/\pi)^{\frac{1}{2}} \tag{3.14}$$

where Q_t is the total loss per unit area, D is the diffusion coefficient of the vapor through soil, C_o is the initial soil concentration, and t the time. Saltzman and Kliger (1979) estimated the diffusion coefficient for DBCP in soil by using eq. 3.14 (Table 3.9). The diffusion coefficient obtained by Call (1957b) for ethylene dibromide in soils varied between 1.38×10^{-3} and 1.38×10^{-4} with a mean value of 6.24×10^{-4}. Since the vapor

TABLE 3.9

Volatilization of DBCP by diffusion through soil (Saltzman and Kliger, 1979)[1]

Soil texture	Solvent[2]	Volatilization ($\mu g/cm^2$)	C (mg/cm^3)	D (cm^2/sec)
Sand	Water	16.97	7.49	5.43×10^{-5}
	Hexane	15.45	7.49	4.50×10^{-5}
Loam	Water	17.99	7.93	5.45×10^{-5}
	Hexane	15.50	7.93	4.04×10^{-5}
Heavy clay	Water	15.60	7.05	5.18×10^{-5}
	Hexane	8.50	7.05	1.54×10^{-5}

[1]D values were estimated from the volatilization loss after 40 hours
[2]DBCP was applied on dry soil in water or hexane

Since the vapor pressure of ethylene dibromide is 11 mm Hg at 25°C, as compared with 0.8 mm Hg at 21°C for DBCP, the values obtained by Saltzman and Kliger (1979) appear reasonable.

Guenzi and Beard (1970) demonstrated the effect of water content on the volatilization of lindane and DDT from soil (Fig. 3.12). They observed that DDT and lindane were lost at a constant rate for each soil during the drying cycle until the soil contained less than a monolayer of water on the soil surface. No further volatilization occurred after the soil reached that degree of dryness. Thus, in the moisture range of 1/3 bar suction to approximately a monolayer, volatilization was independent of water content. These findings were in agreement with those reported by Spencer et al. (1969) and Spencer and Cliath (1970a,b) for dieldrin and lindane.

Temperature effects the volatilization of pesticides from soils by a direct influence on the vapor pressure of pesticides and the physical and chemical properties of the soil. Increase in temperature results in an increase in the volatilization rate of a pesticide in soil (Fang et al., 1961; Harris and Lichtenstein, 1961; Kearney et al., 1964; Gray and Weierch, 1965; Parochetti and Warren, 1966; Guenzi and Beard, 1970; Farmer et al., 1972). Temperature also may effect volatilization of a pesticide through its

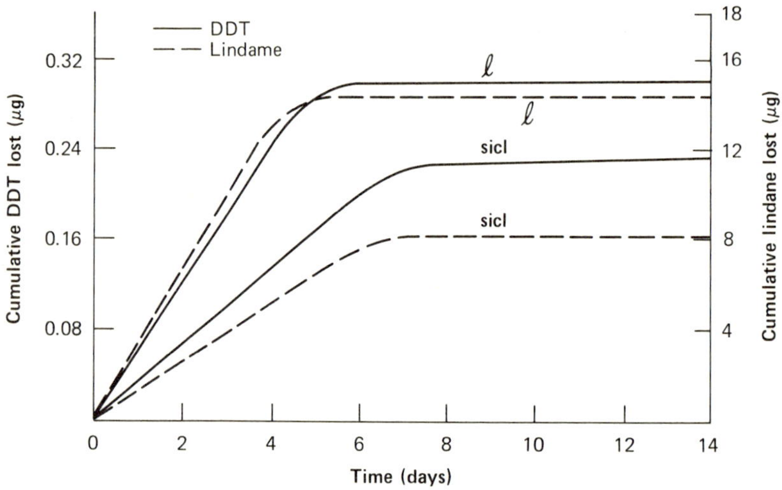

Fig. 3.12. DDT and lindane volatilization from soil during one dry cycle at 30°C (Guenzi and Beard, 1970). ℓ = loam soil, sicl = silty clay loam soil.

effect on movement of the pesticide on the surface by diffusion or by mass flow in the evaporating water. In the absence of evaporating water, volatilization will occur due to the movement of pesticide to the soil surface by diffusion. However, when water evaporates from the soil surface, an appreciable upward movement of water results in order to replace that evaporated water. Thus, pesticide in the soil solution will move towards the surface by mass flow with evaporating water. In general, both mechanisms operate together in the field where water and pesticides vaporize at the same time.

Volatilization of pesticides may be influenced directly or indirectly by the rate of air flow. More volatilization of chlorinated insecticides with increased air flow rate has been observed (Harris and Lichtenstein, 1961; Farmer et al., 1972; Igue et al., 1972). Farmer et al. (1972) demonstrated that a considerable increase in volatilization of dieldrin occurred by increasing the air flow rate (100% relative humidity) over the wet soil (10% soil water). Table 3.10 shows the potential loss of lindane, dieldrin and DDT from soil by volatilization at 10% soil water content and 100% relative humidity (Farmer et al., 1972). It can be seen that the volatilization rate of each insecticide increases

TABLE 3.10

Potential volatilization of lindane, dieldrin, and DDT from a silt loam soil at 10% soil water content, 100% relative humidity and 30°C (Farmer et al., 1972)[1]

Soil concentration (μg/g)	Air flow (miles/hour)	Volatility (kg/ha/year)		
		Lindane	Dieldrin	DDT
1	0.005	-	0.69	-
	0.018	3.3	1.4	0.28
5	0.005	-	3.8	-
	0.018	19.0	8.9	1.3
10	0.005	-	8.7	-
	0.018	43.2	14.2	2.9
50	0.005	-	15.2	-
	0.018	201.6	21.9	4.7

[1]Based on volatilization rated during the first 24 hour period.

as soil concentration and air flow rate increases.

Vaporization of pesticide degradation products may also be an important pathway for dissipation of pesticides from soil. Degradation products of DDT and lindane are much more volatile than the parent compound (Spencer et al., 1973). Field measurements of atmospheric concentrations of various DDT compounds indicated that more than 60% of the material was *p,p'*-DDE (Spencer et al., 1974).

The reader is referred to a comprehensive review of pesticide volatilization by Spencer et al. (1973).

3.4. CHEMICAL CONVERSION AND DEGRADATION

Chemical conversion and degradation of pesticides in soil is a widespread phenomena that plays an important role in the dissipation of many pesticides in soil. Most of the reactions are mediated through water functioning as a reaction medium, as reactant or both. Chemical degradation of pesticides by hydrolysis and oxidation is quite a common process. Other reactions including chemical reduction or isomerization are important for certain compounds. Nucleophilic substitution reactions, other than hydrolysis, may take place with reactants dissolved in water or with reacting groups of soil organic matter. Reaction with free radicals in soil is also a distinct possibility. Chemical degradation of pesticides that occur in soil may be catalyzed in several different ways. Catalysis by clay surfaces, metal oxides, metal ions, and organic matter have been reported. This section will present a review of the major types of chemical reactions contributing to pesticide degradation in soil environment.

3.4.1. Hydrolysis

3.4.1.1. Insecticides

The chemical hydrolysis of many organophosphorus pesticides in soil is an important step in their degradation. Organophosphorus compounds characteristically undergo alkaline hydrolysis that result in the detoxication of these pesticides. Furthermore, their susceptibility to alkaline hydrolysis is related to their biological activity.

The degradation of diazinon, malathion, and ciodrin proceeds by chemical hydrolysis (Konrad et al., 1967; Konrad and Chesters, 1969; Konrad et al., 1969). Malathion and ciodrin are base hydrolyzed whereas diazinon is acid hydrolyzed. Konrad et al. (1967) provided evidence for the chemical hydrolysis of diazinon by comparing the degradation in soil and soil free aqueous system. Comparison of the products of diazinon hydrolysis in acidic soil free systems with products of diazinon degradation in soil systems showed that the products of hydrolysis were the same. This suggests that hydrolysis is the mechanism of chemical degradation of diazinon (71) in soil (*Scheme 3.2*).

Diazinon degrades equally in autoclaved and nonautoclaved soils (Getzin, 1968). The degradation is enhanced by an increase in temperature, soil moisture content and at lower pH. In sterilized soils, disappearance of diazinon is rapid in an acid clay. Rates of chemical hydrolysis of malathion (76) and ciodrin (67) in soil are more rapid than in soil free systems of comparable pH. The chemical degradation in soil proceeds as shown in *Schemes 3.3 and 3.4* (Konrad and Chesters, 1969; Kondrad et al., 1969).

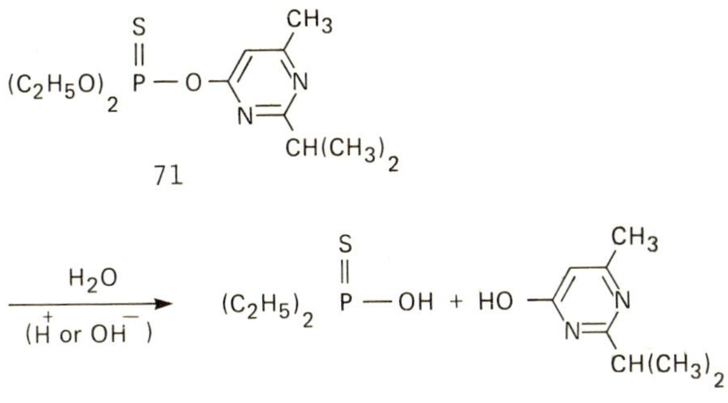

71

126

Scheme 3.2

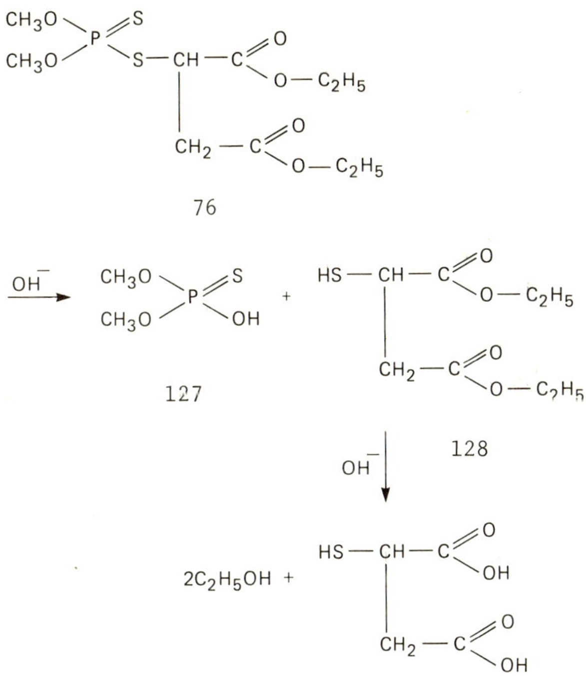

76

127

128

129

Scheme 3.3

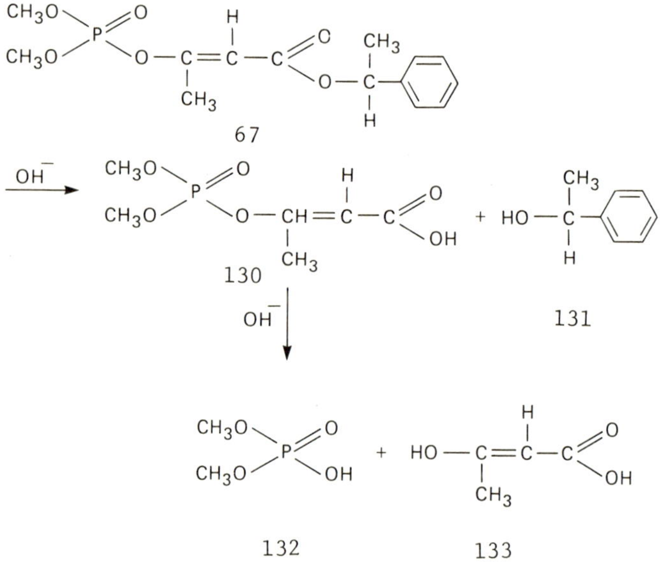

Scheme 3.4

Malaoxon, a degradation product of malathion, is also decomposed in soil by chemical hydrolysis (Paschal and Neville, 1976).

The sorption of diazinon through complexation by exchangeable cations on the soil colloid may also be a mechanism of sorption catalysed hydrolysis (Mortland and Raman, 1967). The ease of Cu^{2+} - catalyzed hydrolysis of several organophosphates was found to be Dursban® > diazinon > ronnel >> Zytron®. Mortland and Raman (1967) postulated that the active molecules undergo coordination with Cu^{2+}, as shown with Dursban® (72) (*Scheme 3.5*). The degradation of ciodrin, as well as of some other organophosphorus insecticides in soil, was also considered to involve sorption catalyzed hydrolysis (Konrad and Chesters, 1969, Konrad et al., 1967, 1969). Susceptibility to acidic or basic hydrolysis may be related to the tendency for sorption catalyzed hydrolysis, since hydrolysis of parathion is apparently not enhanced by the presence of soil (Graetz et al., 1970) and hydrolysis of parathion is less rapid than hydrolysis of diazinion, malathion and ciodrin in the range

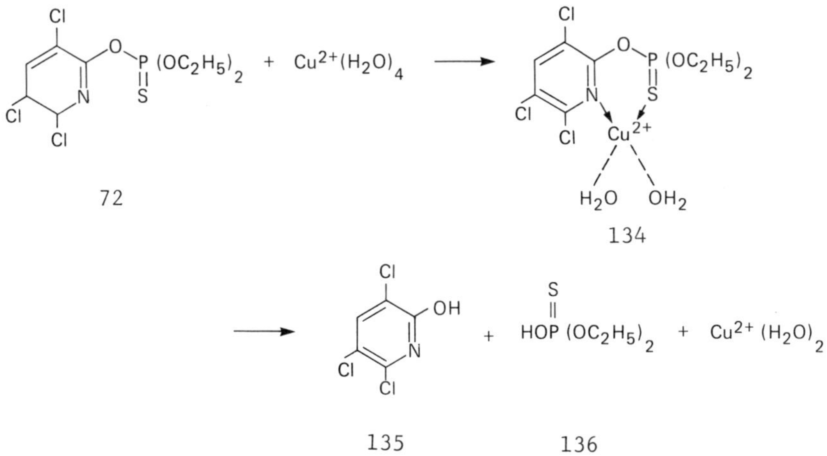

Scheme 3.5

of pH 2 to 9 (Cowart et al., 1971; Gomaa and Faust, 1971).

The chemically induced conversion of a variety of organophosphorus insecticides by clay minerals has been recently established (Mingelgrin and Yaron, 1974; Yaron, 1975; Prost et al., 1976; Mingelgrin et al., 1977; Yaron and Saltzman, 1978). Degradation proceeds by the hydrolysis of P-XA bond (where X is O or S, and A is the electron attracting moiety of the organic molecule). The 1:1 type of clay (kaolinite) enhances a direct hydrolysis of the parathion (69). However, the 1:2 type of clay (bentonite) favors the degradation through a molecular rearrangement prior to hydrolysis (*Scheme 3.6*) as determined by differential infrared spectroscopy (Mingelgrin et al., 1978). The decomposition rates of parathion in kaolinitic and montmorillonitic soils, with similar amounts of clay and organic matter were found to be different (Yaron, 1975). The decomposition in kaolinitic soil was greater than in montmorillonitic soil (Fig. 3.13). Ca-kaolinite clay surfaces exert a strong catalytic effect on parathion degradation, whereas other clays exert only a weak effect. The catalytic

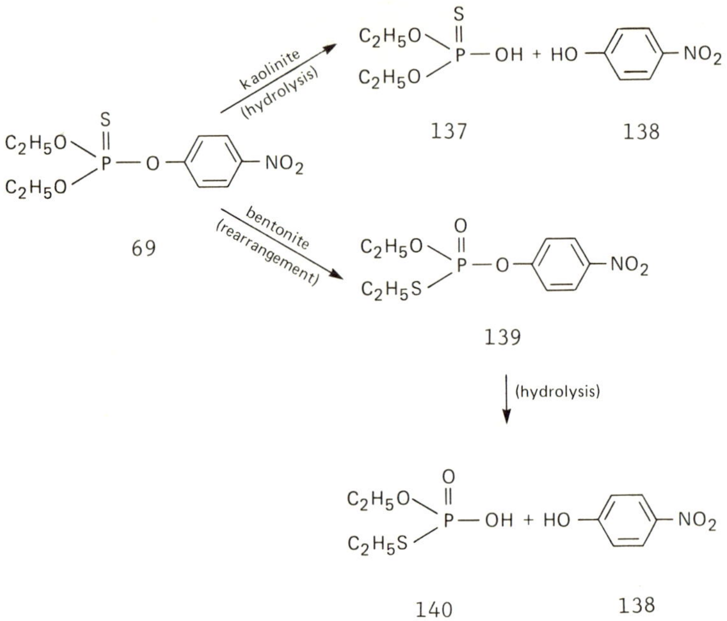

Scheme 3.6

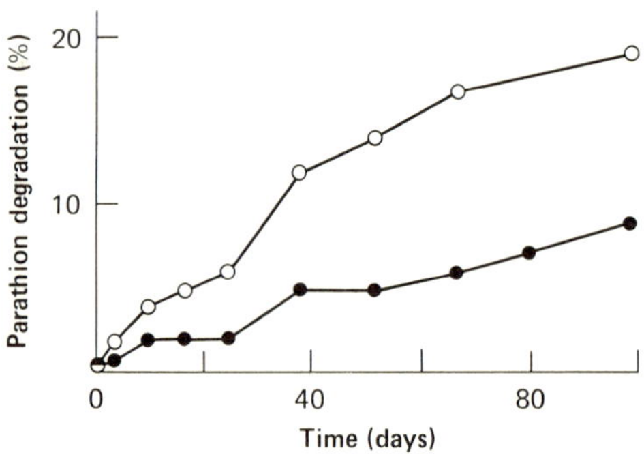

Fig. 3.13. Percentage of water soluble degradation products of parathion recovered from a kaolinitic and montmorillonitic soil during 100 days of incubation at room temperature: • = montmorillonitic and o = kaolinitic (Yaron, 1975). Published by the permission of Springer-Verlag, New York.

effect of kaolinite on the hydrolysis of parathion is highly moisture dependent and the water molecules associated with the exchangeable cations participates in the hydrolysis (Saltzman et al., 1976). The chemically induced hydrolysis of parathion on kaolinite occurs through the attack of a water molecule of an exchangeable cation on the phosphate ester bond (Yaron and Saltzman, 1978). A procedure involving thin layer chromatography, gas chromatography and U.V. spectroscopy was developed to demonstrate the hydrolysis on kaolinite surfaces for a number of phosphoric and phosphorothioic esters (Mingelgrin et al., 1979). The degradation occurs in two first order stages: the first, very fast and short lived, and the second, slower and continuous (Fig. 3.14). In the first stage,

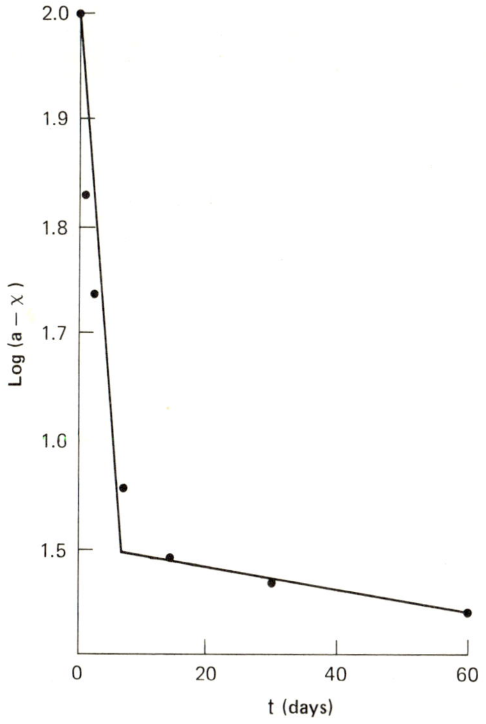

Fig. 3.14. Kinetics of parathion hydrolysis on a dry Ca^{2+} - kaolinite at 22°C; a is the inital amount and x is the amount hydrolyzed at time t (Saltzman et al., 1974). Published by the permission of the Soil Science Society of America.

the parathion molecules specifically adsorbed at the saturating cations are quickly hydrolyzed by contact with the dissociated hydration water molecules. In the second stage, the parathion molecules that have been initially bound to the clay surface by different mechanisms are hydrolyzed when they reach active sites in a proper orientation (Saltzman et al., 1974).

Rao and Sethunathan (1979) observed that an addition of ferrous sulfate to flooded soil led to more rapid and extensive degradation of parathion. This was partly attributed to a low reduction potential under flooded conditions.

Compounds that are highly retained by the soil matrix are often resistant to degradation even though inherently labile (Furmidge and Osgerby, 1967). When hydrolysis appears to be the major degradative pathway, this behavior is likely to be the case for those chemicals with low water solubility (Freed et al., 1979). Thus, when microbial and chemical degradations are relatively slow, compounds that are easily hydrolyzed in water may become much more persistent when incorporated into soil. Adsorption effectively transfers a proportion of a chemical from the aqueous environment to the soil medium. If the soil surface is relatively inert, the adsorption process will have a net effect of protecting the adsorbed species from hydrolysis (Furmidge and Osgerby, 1967). On the other hand, if the soil surfaces are highly reactive, the adsorbed species may become even more susceptible to degradation depending on the type of pesticide and soil properties. For those compounds not readily susceptible to hydrolysis, adsorption does little to increase persistence (Freed et al., 1979).

3.4.1.2. Herbicides

The chemical hydrolysis of *s*-triazines plays an important role in the degradation of these herbicides in soil. Armstrong et al. (1967) observed the formation of hydroxyatrazine as the degradation product of atrazine in soil perfusion columns. Atrazine hydrolysis occurred in sterilized soil at a pH of 3.9. The hydrolysis rate was tenfold greater in the presence of the soil than in its absence at the same pH, thus indicating that atrazine hydrolysis was catalyzed by contact with soil. Harris (1967b) provided evidence for the partial conversion of atrazine, simazine and propazine to their hydroxy derivatives during incubation in soils at 30°C for

8 weeks. The amounts of hydroxy derivatives formed were not affected by the addition of 200 ppm sodium azide as a microbial inhibitor.

Recently, Skipper et al. (1978) provided infrared evidence for the hydrolysis of atrazine on soil colloids. Infrared absorption bands characteristic of the atrazine molecule are the triazine ring out-of-plane deformation at 806 cm^{-1} and skeletal νC=N bands at 1555 cm^{-1} to 1580 cm^{-1} (broad) and 1622 cm^{-1} (Fig. 3.15). The

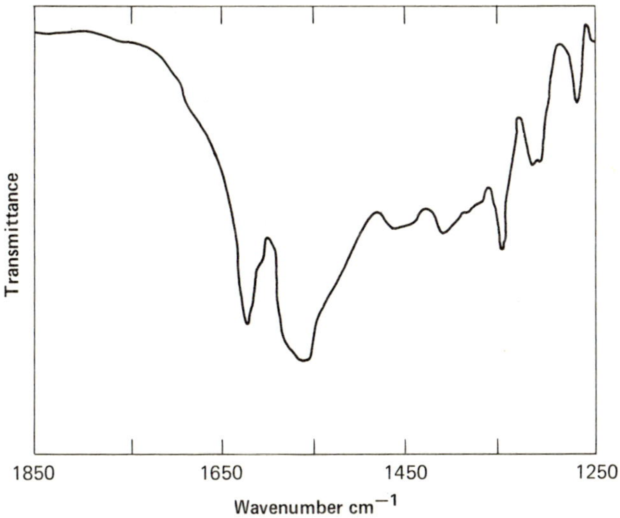

Fig. 3.15. Infrared spectrum of atrazine (Skipper et al., 1978).

interaction of atrazine with H^+- or Al^{3+}- montmorillonite results in a strong carbonyl band at 1745 cm^{-1} demonstrating the formation of hydroxyatrazine from the hydrolysis of atrazine (Fig. 3.16).

A number of factors affect the rate of hydrolysis of *s*-triazines in soils. The pH of the soil and organic matter content largely control the rate of atrazine hydrolysis. In general, the rate is greater in soils containing high organic matter content and low pH (Armstrong et al., 1967). The mechanism of soil catalysis appears to be directly related to the extent of atrazine adsorption (Armstrong and Chester, 1968). Nearpass (1972) reported that chemical hydrolysis of propazine was catalyzed by adsorption on soil organic matter. Brown and White (1969) showed that montmorillonite was the most effective mineral in the hydrolysis of 12 *s*-triazine herbicides by soil clays. Thompson (1968) observed that adsorption of 2-chloro-*s*-triazines onto H^+-HA was accompanied by

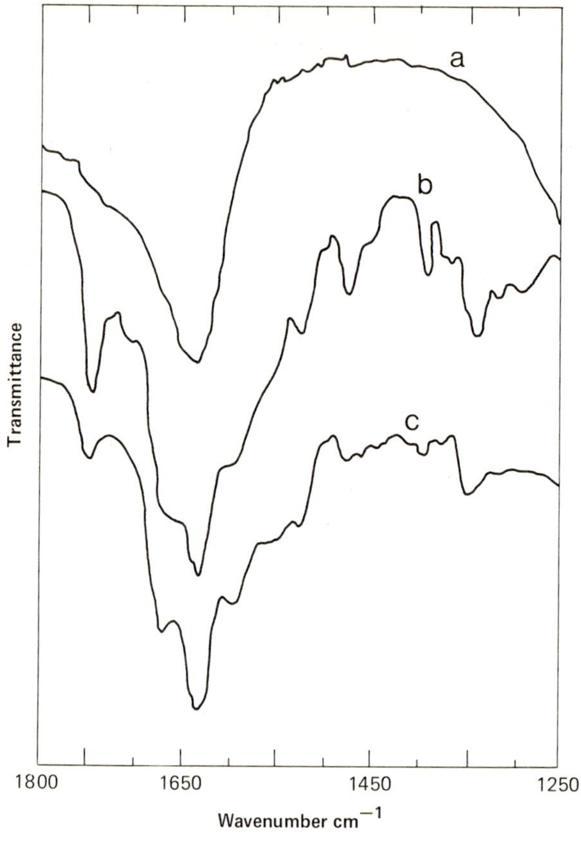

Fig. 3.16. Infrared spectra of (a) Al^{3+}-montmorillonite and the products of atrazine reacted with (b) H^+-montmorillonite, and (c) Al^{3+}-montmorillonite (Skipper et al., 1978).

hydrolysis at 70°C. However, very little hydrolysis was observed at room temperature. Russell et al. (1968b), and Brown and White (1969) provided spectroscopic evidence indicating that montmorillonite clay causes the protonation and subsequent hydrolysis of 2-chloro-*s*-triazines. Infrared spectral data suggests the presence of the keto form of the hydroxy analogue of atrazine and propazine (Russell et al., 1968). Thus, two tautomeric forms of the hydroxy analogues are possible in which the keto form (141) predominates in the protonated hydroxy species. Structure 144 was suggested to be the most likely form of the adsorbed, protonated hydroxytriazines. *Scheme 3.7* shows unprotonated (141, 142) and some

Keto | Enol

141 | 142

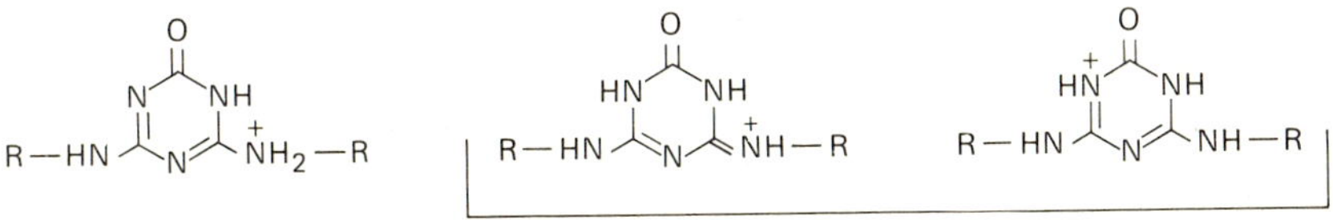

143 | 144

Scheme 3.7

possible tautomeric (143) and resonance (144) structures of protonated hydroxy analogues of chloro-*s*-triazines (Russell et al., 1968a; Skipper et al., 1978). Cruz et al. (1968) observed that the adsorption of propazine and prometone by montmorillonite was followed by protonation and hydroxylation on the montmorillonite surface. The H^+- and Al^{3+}- saturated montmorillonite promoted atrazine hydrolysis whereas Ca^{2+}- or Cu^{2+}- saturated montmorillonite did not (Skipper et al. 1978).

The participation of soil organic matter fractions in atrazine degradation has been demonstrated by several workers. The rate of hydrolysis in aqueous suspension of HA at pH 4 was first order in relation to atrazine concentration. The half life of atrazine, resulting from the first order plot, varied nonlinearly with the concentration of HA (Li and Felbeck, 1972b). Recently, Khan (1978) investigated the kinetics of hydrolysis of atrazine in aqueous FA solution. The logrithm of atrazine concentration was plotted against time in accordance with the following rate equation:

$$-\frac{dC}{dt} = K_{ob}C \qquad (3.15)$$

where C is the residual concentration of atrazine at time t and K_{ob} is the observed rate constant (t^{-1}). It is assumed that the amounts of water or FA consumed in the course of reaction were considered negligible and their concentrations were regarded constant. Linear curves were obtained thereby indicating that atrazine hydrolysis in aqueous FA solution follows first order reaction kinetics with respect to the herbicide concentration (Fig. 3.17).

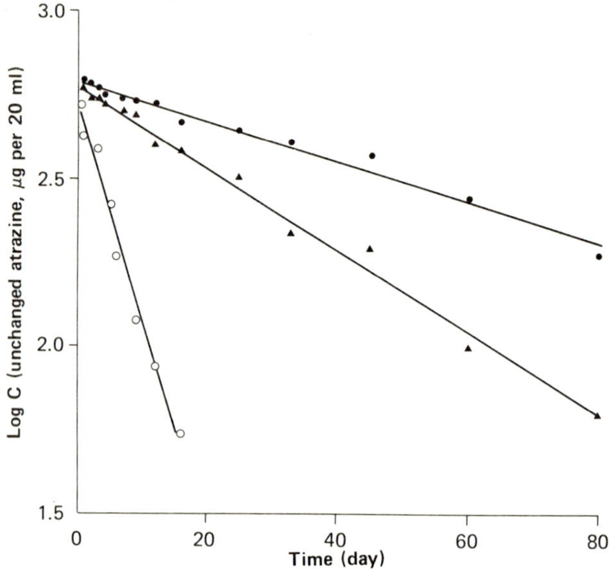

Fig. 3.17. Hydrolysis of atrazine in aqueous fulvic acid (FA) solution at 25°C. ● = 0.5 mg FA/ml, pH 2.9: ▲ = 1.0 mg FA/ml, pH 2.8; o= 5.0 mg FA/ml, pH 2.4 (Khan, 1978).

Table 3.11 shows the first order hydrolysis rate constants and half lives of atrazine at 25°C. The rate of hydrolysis of atrazine increases with an increase in the amount of FA in solution. This, in turn, leads to a shortening of the half life of atrazine. The half life values are lowest at low pH and increase with increasing pH of the solution (Table 3.11). Li and Felbeck (1972b) observed that the half life of atrazine in a 2% HA suspension at 25°C and at pH 4.0 was 1.73 days. For the hydrolysis of atrazine in pH 3.9 and 4.0 aqueous systems at 25°C, half life values of 309 (Armstrong et al., 1967) and 244 days (Li and Felback, 1972b), respectively, have been reported.

The data obtained for the hydrolysis of atrazine in aqueous FA solution at 25°, 40° and 60°C conform to the Arrhenius equation

TABLE 3.11

Rate constants and half lives of hydrolysis of atrazine at 25°C in aqueous fulvic acid (Khan, 1978)

Concentration of fulvic acid (mg/ml)	pH	Rate constant ($10^3 (K_{ob})$ day^{-1})	Half life ($t_{\frac{1}{2}}$ day)
0.5	2.9	19.9	34.8
	4.5	3.99	174
	6.0	1.74	398
	7.0	0.934	742
1.0	2.8	28.4	24.4
	4.5	12.6	55.0
	6.0	3.16	219
	7.0	1.23	563
5.0	2.4	1.51	4.6
	4.5	43.7	15.9
	6.0	13.2	52.5
	7.0	7.93	87.3

as evidenced by the linear relationships obtained by plotting log (rate constants) against the reciprocal of the absolute temperature (Fig. 3.18). The activation energy of the hydrolysis reaction was calculated from the Arrhenius equation:

$$K_{ob} = A_{ob} \exp (-E_{ob}/RT) \tag{3.16}$$

where K_{ob} is the rate constant (t^{-1}), A_{ob} is a constant referred to as the frequency factor, E_{ob} is the observed activation energy of reaction (kJ mole^{-1}), T the absolute temperature and R is the gas constant. The activation energy requirement for the hydrolysis of atrazine appears to increase with an increase in pH of the FA solution. However, a change in the concentration of FA in solution does not affect the E_{ob} values for hydrolysis (Table 3.12).

Khan (1978) suggested that FA will enhance hydrolysis of atrazine in aqueous solution. The mechanism of reaction may be similar to that for H^+ ion catalyzed hydrolysis of *s*-triazines (Horrobin, 1963). FA has various distinct types of acidic functional groups, such as COOH plus phenolic-and/or enolic OH groups (Schnitzer and Khan, 1972). The degree of ionization of these groups will be governed by the pH of the system. Thus, for example, in the pH range of about 5 to 6 the Type I carboxyl groups, which are ortho to phenolic OH

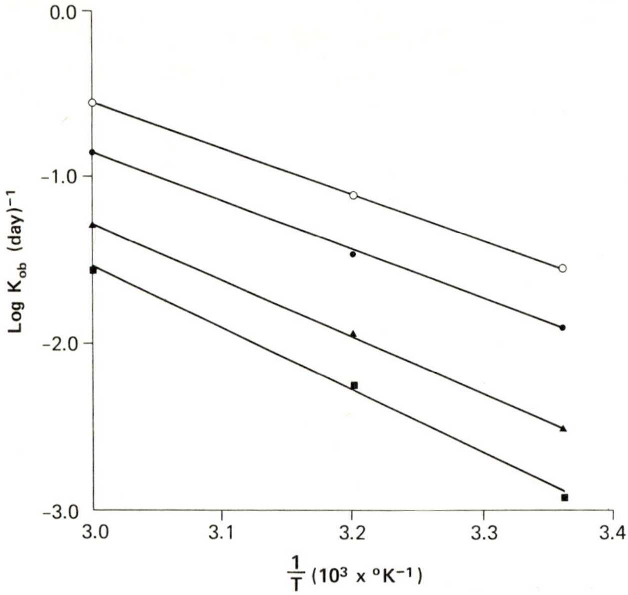

Fig. 3.18. Arrhenius plots for atrazine hydrolysis in aqueous fulvic acid (FA) solution (1.0 mg FA/ml): o = pH 2.8; ● = pH 4.5; ▲ = pH 6.0; ■= pH 7.0 (Khan, 1978).

TABLE 3.12

Activation energy of the hydrolysis of atrazine in aqueous fulvic acid (Khan, 1978)

pH	Activation energy (kJ mol^{-1})		
	0.5 mg fulvic acid/ml	1.0 mg fulvic acid/ml	5.0 mg fulvic acid/ml
2.9	53.6	53.1[1]	50.6[2]
4.5	57.3	56.5	54.4
6.0	63.6	64.9	60.2
7.0	68.6	71.5	70.3

[1]pH 2.8
[2]pH 2.4

groups, are essentially all ionized and Type A acidic functional groups are more than 80% ionized (Gamble, 1972). The latter also include the Type I carboxyl groups and are of greatest chemical interest because they are strongly acidic (Gamble, 1972). This implies that a change in pH of the FA solution would change the types and concentration of acidic functional groups involved in the hydrolysis of atrazine. This in turn may affect the mechanism of hydrolysis as indicated by the change in the activation energy (Table 3.12).

The herbicide pronamide (145) undergoes hydrolysis after cyclization (Yih et al., 1970). Increase of soil temperature results in an increase of reaction rate. The latter also varied widely among soils (*Scheme 3.8*).

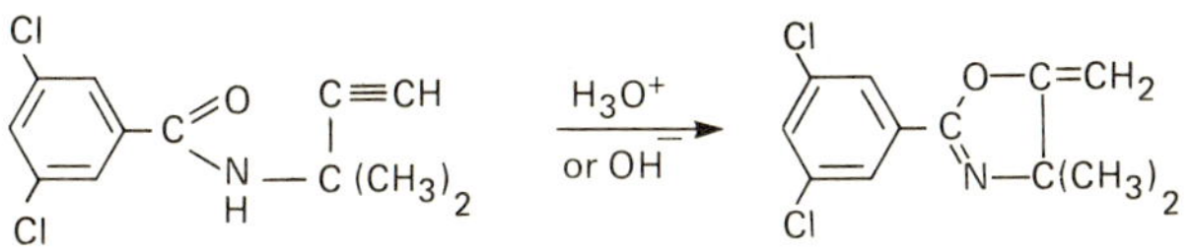

145 146

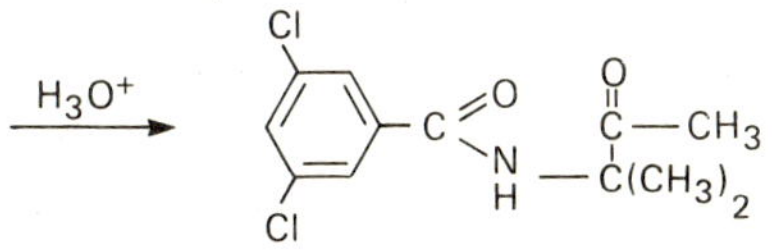

147

Scheme 3.8

Hance (1969c) observed that hydrolysis of atrazine, chlorpropham, diuron, and linuron increased as the soil : solution ratio increased. In acid soils, the herbicides sesone (148) and 2,4-DEP (149) are hydrolyzed to a common intermediate 2,4-dichlorophenoxyethanol (150), which in turn can be biologically oxidized to the herbicide 2,4-D (9) (Carroll, 1952; Viltos, 1952, 1953; Audus, (1952) (*Scheme 3.9*).

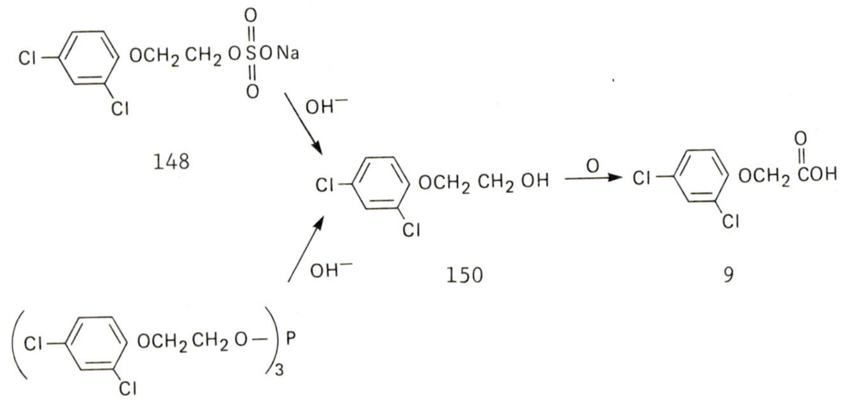

Scheme 3.9

3.4.1.3. *Other pesticides*

Lindane has been shown to be rapidly hydrolyzed in two moist soils (Menn et al., 1965). Castro and Belser (1966) have shown that (Z)- and (E)-1,3-dichloropropene nematicides are hydrolyzed up to 3-fold faster in moist soil than in solution. Chemical conversion of heptachlor to 1-hydroxychlordene is considered one pathway for the insecticide loss in soils (Bowman et al., 1964).

3.4.2. Oxidation and Reduction

Many sulfur containing pesticides are modified in soils by oxidation. Carboxin, a systemic fungicide, is converted to its sulfoxide in autoclaved soil without further reaction (Chin et al., 1970). Parathion can be oxidized to paraoxon (Faust and Suffet, 1966), but this reaction is considered unimportant in soil (Lichtenstein and Schulz, 1964). The epoxidation of aldrin to form dieldrin occurs through chemical reaction (Decker et al., 1965; Edwards, 1966). DDT is reduced to DDD in soils (Guenzi and Beard, 1967). Other examples of nonbiological oxidation of pesticides in soils include decomposition of 3-aminotriazole

(Burchfield and Schechtman, 1958) and the *S*-oxidation of phorate (Getzin and Chapman, 1966).

3.4.3. *N*-Nitrosation

The *N*-nitroso compounds are among the most objectionable substances consumed by man and animals. In recent years, greater attention has been given to nitrogenous pesticides and the possibility of their nitrosation in soil. The reaction calls for favorable pH conditions (about 3-4) and excess nitrite, which is usually lacking in most soils. Under field conditions, the nitrosable residues are usually present in traces and only small quantities of these will actually be nitrosated (IUPAC Special Report, 1977). Production of some *N*-nitrosamines in a soil environment have been shown to result from the interaction of nitrite with agricultural chemicals (Ayanaba et al., 1973; Tate and Alexander, 1974. The *N*-nitrosamine that form may be the *N*-nitroso derivative of the parent compound or a carcinogenic *N*-nitrosamine, such as *N*-nitrosodimethylamine, arising from chemical modification of the pesticide (Ayanaba and Alexander, 1974). Incubation of soil samples amended with NO_2- and dimethylamine showed the formation of *N*-nitrosoamine (Pancholy, 1978). Mills and Alexander (1976) demonstrated that *N*-nitrosodimethylamine was formed in similar quantities in sterilized and nonsterile soil samples thereby suggesting a chemical reaction. In contrast, Oliver et al. (1979) suggested that degradation of certain herbicide related *N*-nitrosamines in aerobic soils was due to microbiological processes.

The formation of *N*-nitrosoatrazine in soil was demonstrated by Kearney et al. (1976). *N*-nitrosoatrazine was detected after one week in soils receiving 2 ppm atrazine and 100 ppm N (as $NaNO_2$) and maintained at pH 4.0, 5.0, 3.5 and 2.5. In a later study, Kearney et al. (1977) found that *N*-nitrosoatrazine was rapidly degraded in aerobic Metapeake loam; only 12% of the added *N*-nitrosoatrazine could be recovered after one month and after 3 and 4 months, the recovery was less than 1%. They suggested that denitrosation back to atrazine was a major degradation pathway. Kearney et al. (1977) concluded that the possibility of *N*-nitrosoatrazine formation seems extremely remote in good agricultural soils (pH 5.0 - 7.0) receiving normal application of atrazine (2 ppm) and even high rates of nitrogen fertilizers (100 ppm N).

Oliver and Kontson (1978) observed that the formation of *N*-nitrosobutralin in soil occurred only when the soil was heavily amended with $NaNO_2$; however, the limited amount of *N*-nitrosobutralin that did form proved to be quite persistent. Thus, in aerobic soil, a significant portion could be extracted after six months.

Khan and Young (1977) observed the formation of *N*-nitrosoglyphosate (154) when different soils were treated with $NaNO_2$ and the herbicide glyphosate (8) at elevated levels (*Scheme 3.10*).

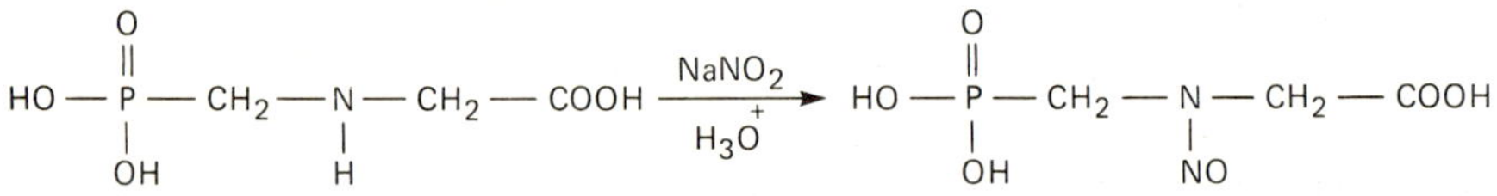

8 154

Scheme 3.10

Although an optimum pH of 2.8-3.0 was found for the formation of *N*-nitrosoglyphosate in solution (Young et al., 1977), pH dependence of the nitrosation of glyphosate in soils of pH range 3.8 to 6.1 was not observed (Khan and Young, 1977). Mills and Alexander (1976) also reported that the amount of dimethylnitrosamine formation in soil was not affected by pH. Khan and Young (1977) observed greater nitrosation in soils with low organic matter and clay content (Table 3.13). Thus in a sandy loam soil about 17 ppm of *N*-nitrosoglyphosate (5.9% theoretical yield) was detected at the end of an 8 day incubation period. The *N*-nitroso derivative formed is persistant in the soil. It was observed that a sandy loam soil treated with 20 ppm nitrite nitrogen and 740 ppm glyphosate contained about 7 ppm of *N*-nitrosoglyphosate even after 140 days (Khan and Young, 1977). It should be recognized that the high levels of the pesticides and $NaNO_2$ employed in the foregoing studies to demonstrate the formation of *N*-nitroso compounds in soil are not likely to be encountered in practical agriculture. For example, the average recommended rates of application of the herbicide glyphosate are about 2 lb/acre. At these levels of

TABLE 3.13

Formation of *N*-nitrosoglyphosate in soils incubated for eight days at 25°C with 20 ppm of nitrite nitrogen as sodium nitrite and 740 ppm glyphosate (Khan and Young, 1977)[1]

Soil texture	Organic matter %	Clay %	Amount of *N*-nitrosoglyphosate formed ppm
Clay	18.0	47.9	ND[2]
Clay loam	4.4	35.0	2.3
Loam	1.1	15.0	5.5
Sandy loam	1.1	5.1	17.1

[1]Distilled water was added to bring the soils to field capacity.
[2]Not detected

application we cannot envisage the formation of *N*-nitrosoglyphosate in soil under normal field conditions.

N-Nitrosodimethylamine is stable in soil (Tate and Alexander, 1975, 1976) and can be translocated from soil into vegetable crops (Dean-Raymond and Alexander, 1976). Dressel (1976) also demonstrated an uptake of *N*-nitrosodimethylamine added to soil by wheat and barley. A recent study by Khan and Marriage (1979) demonstrated that *N*-nitrosoglyphosate can be assimilated by the roots of oat plants and translocated to the shoots. It was observed that *N*-nitrosoglyphosate is not strongly retained by the soil but moved more readily into the root and shoot of oat plants than glyphosate (Table 3.14). It should be realized, however, that under normal field conditions the formation of *N*-nitrosoglyphosate at the levels used by Khan and Marriage (1979) are not expected. Higher concentrations of the herbicide glyphosate and nitrite are essential to get measurable amounts of *N*-nitrosoglyphosate in soil. Even though soil concentrations were extemely high, the observation that *N*-nitrosoglyphosate can be taken up by plants should stimulate further research to determine whether such a possible hazard is in fact a reality with pesticides.

TABLE 3.14

Residue (ppmw on fresh weight basis) of glyphosate and *N*-nitrosoglyphosate in roots and shoots of oat plants grown in the treated soil (Khan and Marriage, 1979)

Treatment (ppmw)	Glyphosate		*N*-nitrosoglyphosate	
	Root	Shoot	Root	Shoot
0	ND[1]	ND	ND	ND
5	ND	ND	4.9	ND
10	ND	ND	9.1	ND
25	4.8	ND	21.3	4.4
50	8.6	1.4	40.3	7.9
100	17.0	3.9	72.7	15.4

[1]Not detected

Since glyphosate is relatively persistent when applied to irrigation water (Comes et al., 1976) and under certain conditions nitrite can accumulate in soil (Chapman and Liebig, 1952) or be constituent in runoff water (Tabatabai, 1974), a possibility for *N*-nitrosoglyphosate formation may exist. Glyphosate is nitrosated by third order kinetics to *N*-nitrosoglyphosate (Young and Khan, 1978). The nitrosation at 25°C is maximum at the reaction pH of 2.5 and has a pH dependent rate constant of 2.43 M^{-2} $sec.^{-1}$. An activation energy of 9.5 kCal $mole^{-1}$ also suggests that glyphosate is nitrosated very readily (Young and Khan, 1978).

N-Nitroso compounds are present in some pesticide formulations that are used extensively in agriculture (Fine et al., 1976). Recently Bontoyan et al. (1979) investigated the extent of *N*-nitrosamine contamination in technical and formulated products used both in agriculture and in or around homes. Of the 91 pesticides and starting materials screened, 25 contained a *N*-nitrosamine at or above 1 ppm. The *N*-nitrosamine found can enter the soil environment through application of the pesticides.

N-Nitrosodipropylamine is a trace contaminant in the herbicide trifluralin (Ross et al., 1977). However, no detectable nitrosamine residues were observed in any crops treated with trifluralin (Sheldon and Day, 1979). Trace quantities of *N*-nitrosodipropylamine resulting from the application of trifluralin can dissipate from soil by volatilization and degradation to volatile and nonvolatile products (Saunders et al., 1979).

3.4.4. Other Reactions

The free radicals in soils may induce pesticide degradation. The reaction with free radicals in soils was considered by Kaufman et al. (1968) and Plimmer et al. (1967) to be responsible for amitrole degradation. Other free radical generating systems also degrade amitrole in vitro (Castelfranco et al., 1963).

Plimmer et al. (1970) reported the identification of 1,3-bis (3',4'-dichlorophenyl)-1,2,3-triazene (153) from soil originally containing 3,4-dichloroaniline (151). They suggest that this compound arises by diazotization of the amine by nitrite derived from fertilizer, followed by coupling with more 3,4-dichloro-aniline (*Scheme 3.11*).

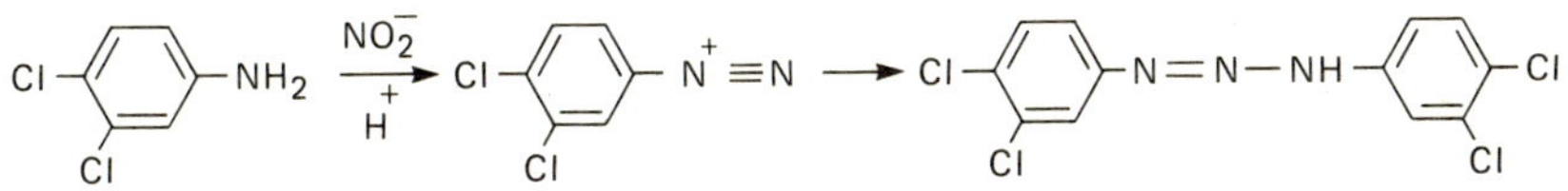

151 152 153

Scheme 3.11

DDT is slowly converted to DDE in sterile soil. Diffusion of DDT through clay minerals results in a considerable amount of DDE as the degradation product (Lopez-Gonzalez and Valenzuele-Calahorro, 1970). Degradation results from the interaction of DDT with active zones on the surface of homoionic clay minerals during diffusion through the pesticide free clay. Furthermore, DDT decomposes more in the homoionic sodium clay than in the corresponding hydrogen clay. The difference is attributed to the higher pH in the sodium system, which shifts the equilibrium between DDT and DDE towards DDE (Lopez-Gonzalez and Valenzuele-Calahorro, 1970). Guenzi and Beard (1975) suggested chemical conversion of DDT to DDE and the conversion was enhanced by increasing temperature. The role of the water in the conversion process is not known but it enhances the process.

3.5 PHOTODECOMPOSITION

Solar radiation is responsible for many chemical changes of pesticides in the environment. Within the range of ultraviolet (UV) sunlight wavelengths (290 to 450 nm), sufficient energy exists to bring about many chemical transformations of pesticides. Often the degradation products are identical with those produced by chemical and biological reactions, however, photodecomposition has produced some unique structures. For photodecomposition, the light with wavelengths in the UV spectrum must come in contact with the pesticide. Since penetration of UV light into solid matter is limited, photodecomposition of pesticides in soil is restricted to residues on or very near the surface. The extent of photodecomposition depends on the duration of exposure, the intensity and wavelength of the light, the state of the chemicals, the nature of the supporting medium or solvent, pH of the solution and the presence of water, air, and photosensitizers.

In this section no attempt is made to include an exhaustive coverage of the photochemistry of pesticides; the reader is directed to several pertinent reviews by Crosby (1976), Moilanen et al. (1975) and Plimmer (1970). The methods used in the study of the photochemical degradation of pesticides have been recently described by Cavell (1979).

Photodecomposition of chemicals has been reported for a wide range of pesticides used in agriculture. However, the role of photochemical reactions in the degradation of pesticides in soil is uncertain as most of these reactions have been reported under conditions involving exposure to high intensity light and frequently in nonaqueous solvents. Photodecomposition may be of considerable importance for pesticides applied to the soil surface.

Photolysis of trifluralin on a soil surface was observed by Wright and Warren (1965), however, no products were identified. Kuwahara et al. (1965) showed that PCP was decomposed in rice field water after several days of exposure to sunlight. Sunlight has been considered as a major factor in the loss of herbicidal activity of organoborates under arid conditions (Rake, 1961). Asai et al. (1969) observed that photolysis of endrin on some air dry soil resulted in the formation of ketoendrin and a related aldehyde. Holmstead et al. (1978) investigated permethrin photodecomposition on a 0.25 mm thickness thin layer plates under sunlight.

Photolysis resulted in cyclopropane ring isomerization and ester cleavage to 3-phenoxybenzyl alcohol and the dichlorovinyl acid. Trace amounts of the esters were also formed. Smith et al. (1978) developed a technique for the production of reproducible thin layers of pesticide containing soil for studies involving residue behavior on air dry soil under different environmental conditions. It was observed that methidathion on a thin layer of dry soil exposed to sunlight produced considerable quantities of methidathion oxygen analogue. Liang and Lichtenstein (1976) examined the effect of soils on photodecomposition of [^{14}C] azinphosmethyl. The air dried soil was treated with [^{14}C] azinphosmethyl and a portion of it was placed in rectangular glass dishes. Exposure to sunlight or UV light for eight hours resulted in the degradation of the insecticide. With increasing soil moisture content, increased degradation occurred with UV light, but not with sunlight (Liang and Lichtenstein, 1976). The photodecomposition of the herbicide basagran was investigated on soil thin layer plates (Nilles and Zabik, 1975). The major routes of phototransformation of this herbicide were found to be oxidative dimerization and a nonconcerted loss of SO_2. Decomposition of eleven dinitroaniline herbicides, applied to dry soil thin layer plates and exposed to sunlight, was higher than if held in the dark under otherwise similar conditions (Parochetti and Dec, 1978). Kennedy and Talbert (1977) observed losses of dinitroaniline herbicides on soil TLC plates when exposed to UV light.

Diazinon, methidathion and profenofos were readily degraded on soil surfaces under artificial sunlight (Burkhard and Guth, 1979). The rate of degradation decreased in the order diazinon, profenofos, methidathion and was always greater in moist than in dry soil. The major photolysis products identified were 2-isopropyl-6-methylpyrimidin-4-ol from diazinon, 5-methoxy-3H-1,3,4-thiadiazol-2-one from methidathion and 4-bromo-2-chlorophenol and 4-bromo-2-chlorophenyl ethyl hydrogen phosphate from profenofos. Burkhard and Guth (1979) observed the formation of same compounds in hydrolysis studies and also upon photodecomposition in aqueous solutions of diazinon and methidathion. Profenofos, however, showed a different photolytic reaction in aqueous systems, forming *O*-(2-chlorophenyl) *O*-ethyl *S*-propyl phosphorothioate.

Lack of appropriate light absorption or photochemical stability in distilled water does not preclude light induced pesticide transformations under natural field conditions (Crosby, 1976). Photosensitizers have been shown to occur in natural water and soil solution which may absorb solar energy and transfer it to the pesticide that would not ordinarily undergo solar transformation. Ethylenethiourea (ETU) in aqueous solution (0.5-50 ppm) was stable to sunlight (Ross and Crosby, 1973). However, ETU decomposition occurs in agricultural drainage waters in sunlight thereby indicating that natural photosensitizers may play an important part in the environmental transformations of zenobiotics. The decomposition of oxamyl when exposed to UV light was investigated in both distilled and river water (Harvey and Han, 1978). In both types of water, oxamyl was converted to the corresponding oximino compound (methyl *N*-hydroxy-*N*'-*N*'-dimethyl-1-thiooxamimidate) at an accelerated rate. The initial hydrolysis product was converted gradually to a material identical to the geometrical isomer of the oximino compound. In addition, small amounts of very polar materials were also formed. Decomposition was more rapid in river water than in distilled water (Table 3.15). These results further suggest that natural photosensitizers may play a role in the breakdown reaction.

TABLE 3.15

Breakdown of oxamyl (1 ppm) in distilled and river water under ultraviolet light (Harvey and Han, 1978)

Exposure time (hours)	Percentage composition			
	Oxamyl	Oximino compound	Isomer of oximino compound	Polar metabolites
	Distilled water			
0	100	0	0	0
48	90	8	0	2
96	79	9	2	10
168	61	18	3	18
240 (Dark)	98	2	0	0
	River water			
0	100	0	0	0
48	1	67	12	20
96	2	51	25	22
168	2	51	24	23
240 (Dark)	98	2	0	0

Because surface waters and soil solutions contain naturally occurring organic materials such as humic substances, which can strongly absorb UV light, environmental photochemistry of pesticides may be strongly influenced by natural photosensitization. The FA content of surface waters may vary from 100 to 500 mg/kg (Schnitzer and Khan, 1972). It is possible that FA could act as a photosensitizer for other nonabsorbing pesticides in soil solutions and surface waters. UV irradiation may bring about the photooxidation of organic matter in water (Gjessing and Gjerdahl, 1970; Gjessing, 1976) and the rate is pH-dependent, increasing with increase in pH (Chen et al., 1978). Photolysis of atrazine

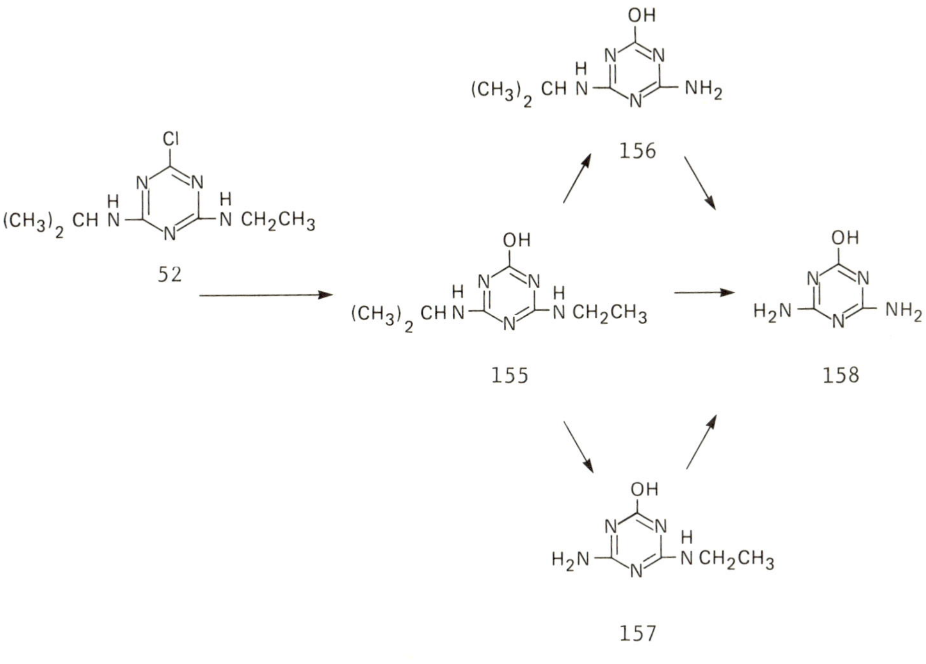

Scheme 3.12

(52) in water yields the 2-hydroxy analogue (155) only (*Scheme 3.12*). However, photolysis under the same conditions in the presence of FA also yields *N*-dealkylated compounds (156) and (157), demonstrating *N*-dealkylation in addition to hydrolysis (Khan and Schnitzer, 1978). Further photochemical *N*-dealkylation of 156 and 157 gives rise to a de-*N*-*N*'-dialkyl analogue, namely, 2-hydroxy-4,6-diamino-*s*-triazine (158). Photolysis of hydroxyatrazine (155) in the presence of FA yields compound 156, 157 and 158 thereby indicating that either FA, or its photoproducts, or both assists successive *N*-dealkylations (Khan and Schnitzer, 1978).

REFERENCES

Aharonson, N. and Kafkafi, U., 1975a. J. Agric. Food Chem., 23: 434-437.

Aharonson, N. and Kafkafi, U., 1975b. J. Agric. Food Chem., 23: 720-724.

Armstrong, D.E. and Chesters, G., 1968. Environ. Sci. Technol., 2: 683-689.

Armstrong, D.E., Chesters, G. and Harris, R.F., 1967. Soil Sci. Soc. Am. Proc., 31: 61-66.

Arnold, J.S. and Farmer, W.J., 1979. Weed Sci., 27: 257-262.

Asai, R.I., Westlake, W.E. and Gunther, F.A., 1969. Bull. Environ. Contam. Toxicol., 4: 278-284.

Ashton, F.M., 1961. Weeds, 9: 612-619.

Ashton, F.M. and Sheets, T.J., 1959. Weeds, 7: 88-90.

Audus, L.J., 1952. Nature (London), 170: 886-887.

Ayanaba, A. and Alexander, M., 1974. J. Environ. Qual., 3: 83-89.

Ayanaba, A., Verstraete, W. and Alexander, M., 1973. Soil Sci. Soc. Am. Proc., 37: 565-568.

Bailey, G.W. and White, J.L., 1970. Residue Rev., 32: 29-92.

Bailey, G.W., White, J.L. and Rothberg, T., 1968. Soil Sci. Soc. Am. Proc., 32: 222-234.

Ballard, T.M., 1971. Soil Sci. Soc. Am. Proc., 35: 145-147.

Barlow, F. and Hadaway, A.B., 1955. Bull. Entomol. Res., 46: 547-559.

Barlow, F. and Hadaway, A.B., 1958. Bull. Entomol. Res., 49: 315-331.

Barnett, A.P., Hauser, E.W., White, A.W. and Holladay, J.H., 1967. Weeds, 15: 133-137.

Bartha, R. and Hsu, T.S., 1976. In: D.D. Kaufman, G.G. Still, G.D. Paulson and S.K. Bandal (Editors), Bound and Conjugated Pesticides Residues, ACS Symp. Ser., 29, pp. 258-271.

Best, J.A., Weber, J.B. and Weed, S.B., 1972. Soil Sci., 114: 444-450.

Biggar, J.W., Mingelgrin, U. and Cheung, M., 1978. J. Agric. Food Chem., 26: 1306-1312.

Bode, L.E., Day, C.L., Gebhardt, M.R. and Goering, C.E., 1973a. Weed Sci. 21: 480-484.

Bode, L.E., Day, C.L., Gebhardt, M.R. and Goering, C.E., 1973b. Weed Sci., 21: 485-489.

Bontoyan, W.R., Law, M.W. and Wright, D.P. Jr., 1979. J. Agric. Food Chem., 27: 631-635.

Bowman, B.T., 1973. Soil Sci. Soc. Am. Proc., 37: 200-207.
Bowman, B.T., 1978. Soil Sci. Soc. Am. J., 42: 441-446.
Bowman, B.T. and Sans, W.W., 1977. Soil Sci. Soc. Am. J., 41: 514-519.
Bowman, B.T., Adams, R.S. Jr. and Feton, S.W., 1970. J. Agric. Food Chem., 18: 723-727.
Bowman, M.C., Schechter, M.S. and Carter, R.L., 1965. J. Agric. Food Chem., 13: 360-365.
Bowman, M.C., Acree, F. Jr., Lofgren, C.S. and Beroza, M., 1964. Science, 146: 1481.
Briggs, G.G. and Dawson, J.E., 1970. J. Agric. Food Chem., 18: 97-99.
Broadbent, F.E. and Bradford, G.R., 1952. Soil Sci., 74: 447-457.
Brown, C.B. and White, J.L., 1969. Soil Sci. Soc. Am. Proc., 33: 863-867.
Burchfield, H.P. and Schechtman, J., 1958. Contr. Boyce Thompson Inst. Pl. Res., 19: 411-416.
Burdon, J., Hayes, M.H.B. and Pick, M.E., 1977. J. Environ. Sci. Health, B12: 37-51.
Burkhard, N. and Guth, J.A., 1979. Pestic. Sci., 10: 313-319.
Burns, R.G., 1972. Proc. Br. Weed Control Congr., 11th, Brighton, pp. 1203-1209.
Burns, R.G. and Audus, L.J., 1970. Weed Res., 10: 49-58.
Burns, I.G. and Hayes, M.H.B., 1974. Residue Rev., 52: 117-146.
Burns, I.G., Hayes, M.H.B. and Stacey, M., 1973a, Weed Res., 13: 79-90.
Burns, I.G., Hayes, M.H.B. and Stacey, M., 1973b. Pestic. Sci. 4: 201-209.
Burnside, O.C., Wicks, G.A. and Fenster, C.R., 1969. Weed Sci., 17: 241-245.
Calderbank, A., 1968. Adv. Pestic. Control Res., 8: 127-235.
Calderbank, A. and Tomlinson, T.E., 1969. PANS, 15: 466-472.
Call, F., 1957a. J. Sci. Food Agric., 8: 86-89.
Call, F., 1957b. J. Sci. Food Agric., 8: 143-150.
Carringer, R.D., Weber, J.B. and Monaco, T.H., 1975. J. Agric. Food Chem., 23: 568-572.
Carroll, R.B., 1952. Contrib. Boyce Thompson Inst., 16: 409-417.
Castelfranco, P., Oppenheim, A. and Yamaguchi, S., 1963. Weeds, 11: 111-115.
Castro, C.E. and Belser, N.O., 1966. J. Agric. Food Chem., 14: 69-70.
Cavell, B.D., 1979. Pestic. Sci., 10: 177-180.
Chang, S.S. and Stritzke, J.F., 1977. Weed Sci., 25: 184-187.
Chapman, H.D. and Liebig, G.M., 1952. Soil Sci. Soc. Am. Proc., 16: 276-282.
Chen, Y., Khan, S.U. and Schnitzer, M., 1978. Soil Sci. Soc. Am. J., 42: 292-296.
Chin, W.T., Stone, G.M. and Smith, A.E., 1970. J. Agric. Food Chem., 18: 731-732.
Choi, J. and Aomine, S., 1972. Soil Sci. Plant Nutr., 18: 255-260.
Choi, J. and Aomine, S., 1974. Soil Sci. Plant Nutri., 20: 135-144.
Chiou, C.T., Freed, V.H., Schmedding, D.W. and Kohnert, R.L., 1977. Environ. Sci. Technol., 11: 475-478.
Chopra, S.L., Das, N. and Das, B., 1970. J. Indian Soc. Soil Sci. 18: 437-446.
Coffey, D.L. and Warren, G.F., 1969. Weed Sci., 17: 16-19.
Cohen, J.M. and Pinkerton, C., 1966. Adv. Chem., Ser. 60: 163.
Comes, R.D., Bruns, V.F. and Kelley, A.D., 1976. Weed Sci., 24: 47-50.
Cowart, R.P., Bonner, F.L. and Epps, E.A. Jr., 1971. Bull. Environ. Contam. Toxicol., 6: 231-234.

Crosby, D.G., 1976. In: P.C. Kearney and D.D. Kaufman (Editors), Herbicides: Chemistry, Degradation and Mode of Action, Dekker, New York, N.Y., pp. 835-890.
Cruz, M., White, J.L. and Russell, J.D., 1968. Israel J. Chem., 6: 315-323.
Damanakis, M., Drennan, D.S.H., Fryer, J.D. and Holley, K., 1970. Weed Res., 10: 264-277.
Dao, T.H. and Lavy, T.L., 1978. Weed Sci., 26: 303-308.
Davidson, J.M. and Chang, R.K., 1972. Soil Sci. Soc. Am. Proc., 36: 257-261.
Davidson, J.M., Rieck, C.E. and Santelmann, P.W., 1968. Soil Sci. Soc. Am. Proc., 32: 629-633.
Dean-Raymond, D. and Alexander, M., 1976. Nature (London) 262: 294-296.
Decker, J.C., Bruce, W.N. and Bigger, J.H., 1965. J. Econ. Entomol., 58: 266-271.
Deli, J. and Warren, G.F., 1971. Weed Sci., 19: 67-69.
Deming, J.M., 1963. Weeds, 11: 91-96.
Dickens, R. and A.E. Hiltbold., 1967. Weeds, 15: 299-304.
Dieguez-Carbonell, D. and Pascual, C.R., 1975. Environ. Qual. Safety III, 237-242.
Dixon, J.B., Moore, D.E., Agrihotn, N.P. and Lewis, D.E., Jr. 1970. Soil Sci. Soc. Am. Proc., 34: 805-808.
Doherty, P.J. and Warren, G.F., 1969. Weed Res., 9: 20-26.
Dowdy, R.H. and Mortland, M.M., 1967. Soil Sci., 105: 36-43.
Dressel, J., 1976. Z. Lebensm. Unters. Forsch., 163: 11-13.
Edwards, C.A., 1966. Residue Rev., 13: 83-132.
Edwards, C.A., Beck, S.D. and Lichtenstein, E.P., 1957. J. Econ. Entomol., 50: 622-626.
Ehlers, W., Letey, J., Spencer, W.F. and Farmer, W.J., 1969. Soil Sci. Soc. Am. Proc., 33: 501-504.
Epstein, E. and Grant, W.J., 1968. Soil Sci. Soc. Am. Proc. 32: 423-426.
Fang, S.C., Thiesen, P. and Freed, V.H., 1961. Weeds, 9: 569-574.
Farmer, W.J. and Jensen, C.R., 1970. Soil Sci. Soc. Am. Proc., 34: 28-31.
Farmer, W.J., Igue, K., Spencer, W.F. and Martin, J.P., 1972. Soil Sci. Soc. Am. Proc., 36: 443-447.
Farmer, W.J., Igue, K., and Spencer, W.F., 1973. J. Environ. Qual., 2: 107-109.
Faust, S.D. and Suffet, I.H., 1966. Residue Rev., 15: 44-116.
Felsot, A. and Dahm, P.A., 1979. J. Agric. Food Chem., 27: 557-563.
Fine, D.H., Ross, R., Fan, T., Rounbehler, D.P., Silvergleid, A., Song, L. and Morrison, J., 1976. Abst. 172nd Am. Chem. Soc. Nat'l. Meeting, San Francisco, California.
Fleming, W.E., 1950. J. Econ. Entomol., 43: 87-89.
Fleming, W.E. and Maines, W.W., 1953. J. Econ. Entomol., 46: 445-449.
Fleming, W.E. and Maines, W.W., 1954. J. Econ. Entomol., 47: 165-169.
Freed, V.H., Chiou, C.T. and Schmedding, D.W., 1979. J. Agric. Food Chem., 27: 706-707.
Frissel, M.J. and Bolt, G.H., 1962. Soil Sci., 94: 284-291.
Furmidge, C.G.L. and Osgerby, J.M., 1967. J. Sci. Food Agric., 18: 269-273.
Gaillardon, P., 1975. Weed Res., 15: 393-399.
Gamar, Y. and Mustafa, M.A., 1975. Soil Sci. 119: 290-295.
Gamble, D.S., 1972. Can. J. Chem., 50: 2680-2690.
Geissbühler, H., Haselback, C. and Aebi, C., 1963. Weed Res., 3: 140-153.

Getzin, L.W., 1968. J. Econ. Entomol., 61: 1560-1565.
Getzin, L.W. and Chapman, R.K., 1959. J. Econ. Entomol., 52: 1160-1165.
Getzin, L.W. and Chapman, R.K., 1960. J. Econ. Entomol., 53: 47-51.
Giles, C.H., MacEwan, T.H., Nakhwa, S.N. and Smith, D., 1960. J. Chem. Soc., pp. 3973-3993.
Gilmour, J.T. and Coleman, N.T., 1971. Soil Sci. Soc. Am. Proc., 35: 256-259.
Gjessing, E.T., 1976. Physical and Chemical Characteristics of Aquatic Humus, Ann Arbor Science, Ann Arbor, Michigan, pp. 85-89.
Gjessing, E.T. and Gjerdahl, T., 1970. Vatten, 12: 144-146.
Gomaa, H.M. and Faust, S.D., 1971. In: S.J. Faust and J.V. Hunter (Editors), Organic Compounds in Aquatic Environments, Dekker, New York, N.Y., pp. 341-376.
Goring, C.A.I., 1967. Ann. Rev. Phytopathol., 5: 285-318.
Graetz, D.A., Chesters, G., Daniel, T.C., Newland, L.W. and Lee, G.B., 1970. J. Water Pollution Fed., 42: R76-R94.
Graham-Bryce, I.J., 1969. J. Sci. Food Agric., 20: 489-494.
Gray, R.A. and Weierich, A.J., 1965. Weeds, 13: 141-147.
Gray, R.A. and Weierich, A.J., 1968. Proc. 9th Br. Weed Contr. Conf., 1: 94-101.
Greenland, D.J., 1965. Soils & Fert., 28: 415-425.
Grice, R.E., Hayes, M.H.B. and Lundie, P.R., 1973. Proc. 7th Br. Insect. Fungicide Conf., pp. 73-81.
Grim, R.E., 1968. Clay Mineralogy, 2nd ed. McGraw-Hill Book Co., New York, N.Y., 596 pp.
Grover, R., 1968. Weed Res., 8: 226-232.
Grover, R., 1971. Weed Sci., 19: 417-418.
Grover, R., 1977. Weed Sci., 25: 159-162.
Guenzi, W.D. and Beard, W.E., 1967. Science, 156: 1116-1117.
Guenzi, W.D. and Beard, W.E., 1967. Soil Sci. Soc. Am. Proc., 31: 664-647.
Guenzi, W.D. and Beard, W.E., 1970. Soil Sci. Soc. Am. Proc., 34: 443-447.
Guenzi, W.D. and Beard, W.E., 1975. Environ. Qual. Safety Suppl., III, 214-221.
Hadzi, D., Klofutar, C. and Oblak, S., 1968. J. Chem. Soc., A, 905-908.
Hamaker, J.W., 1972. In: C.A.I. Goring and J.W. Hamaker (Editors), Organic Chemicals in the Soil Environment, Vol 1, Dekker, New York, N.Y., pp. 341-397.
Hamaker, J.W. and Thompson, J.M., 1972. In: C.A.I. Goring and J.W. Hamaker (Editors), Organic Chemicals in the Soil Environment, Vol. 1, Dekker, New York, N.Y., pp. 49-143.
Hamaker, J.W., Goring, C.A.I. and Youngson, C.R., 1966. Advan. Chem. Ser., 66: 23-27.
Hance, R.J., 1965a. Weed Res., 5: 98-107.
Hance, R.J., 1965b. Weed Res., 5: 108-114.
Hance, R.J., 1967. Weed Res., 7: 29-36.
Hance, R.J., 1969a. Can. J. Soil Sci., 49: 357-364.
Hance, R.J., 1969b. Weed Res., 9: 108-113.
Hance, R.J., 1969c. J. Sci. Food Agric., 20: 144-145.
Hance, R.J., 1971. Weed Res., 11: 106-110.
Haque, R. and Sexton, R., 1968. J. Colloid Interface Sci., 27: 818-827.
Haque, R. and Coshow, R.W., 1971. Environ. Sci. Technol., 5: 138-141.

Haque, R., Lilley, S. and Coshow, W.R., 1970. J. Colloid Int. Sci., 33: 185-188.
Harris, C.I., 1966. Weeds, 14: 6-10.
Harris, C.I., 1967a. Weeds, 15: 214-216.
Harris, C.I., 1967b. J. Agric. Food Chem., 15: 157-162.
Harris, C.I., 1969. J. Agric. Food Chem., 17: 80-82.
Harris, C.I. and Sheets, T.J., 1965. Weeds 13: 215-219.
Harris, C.I. and Warren, G.F., 1964. Weeds, 12: 120-126.
Harris, C.I. and Lichtenstein, E.P., 1961. J. Econ. Entomol., 54: 1038-1045.
Harris, C.R. and Bowman, B.T., 1976. Soil Sci. Soc. Am. J., 40: 385-389.
Hartley, G.S., 1969. Advan. Chem., Series 86: 115-134.
Harvey, J. Jr. and Han, J.C.Y., 1978. J. Agric. Food Chem., 26: 536-541.
Harvey, R.G., 1974. Weed Sci., 22: 120-124.
Hayes, M.H.B., 1970. Residue Rev., 32: 131-174.
Hayes, M.H.B., Stacey, M. and Thompson, J.M., 1968. In: Isotopes and Radiation in Soil Organic Matter Studies, I.A.E.A., Vienna, pp. 75-90.
Hayes, M.H.B., Pick, M.E. and Toms, B.A. 1975. Residue Rev., 57: 1-25.
Hayes, M.H.B., Pick, M.E., Stacey, M. and Toms, B.A., 1972. Proc. Internat. Clay confr., Madrid, p. 675.
Hayes, M.H.B., Pick, M.E., Stacey, M. Toms, B.A. and Quinn, C.M., 1974. Proc. xth Internat. Congr. Soil Sci., Moscow 7: p. 90.
Helling, C.S., 1970. Residue Rev., 32: 175-210.
Helling, C.S., 1971a. Soil Sci. Soc. Am. Proc., 35: 732-737.
Helling, C.S., 1971b. Soil Sci. Soc. Am. Proc., 35: 737-743.
Helling, C.S., 1971c. Soil Sci. Soc. Am. Proc., 35: 743-748.
Helling, C.S. and Turner, B.C., 1968. Science, 162: 562-563.
Helling, C.S., Kearney, P.C. and Alexander, M., 1971. Adv. Agronomy 23: 147-240.
Herr, D.E., Stronbe, E.W. and Ray, D.A., 1966. Weeds, 14: 248-250.
Hilton, H.W. and Yuen, Q.H., 1963. J. Agric. Food Chem., 11: 230-234.
Hilton, H.W. and Yuen, Q.H., 1966. J. Agric. Food Chem., 14: 86-90.
Hindin, E., May, D.S. and Dunstan, G.H., 1966. Advan. Chem., Ser., 60: 132-145.
Hollist, R.L. and Foy, C.L., 1971. Weed Sci., 19: 11-16.
Holmstead, R.L., Casida, J.E., Ruzo, L.O. and Fullmer, D.G., 1978. J. Agric. Food Chem., 590-595.
Hornsby, A.G. and Davidson, J.M., 1973. Soil Sci. Soc. Am. Proc., 37: 823-828.
Horrobin, S., 1963. J. Chem. Soc., p. 4130.
Hsu, T.S. and Bartha, R., 1974a. Soil Sci., 116: 444-452.
Hsu, T.S. and Bartha, R., 1974b. Soil Sci., 118: 213-220.
Hsu, T.S. and Bartha, R., 1976. J. Agric. Food Chem., 24: 119-122.
Huggenberger, F.J., Letey, J. and Farmer, W.J., 1972. Soil Sci. Soc. Am. Proc., 36: 544-548.
Igue, K., Farmer, J., Spencer, W.F. and Martin, J.P., 1972. Soil Sci. Sco. Amer. Proc., 36: 447-450.
IUPAC, Pesticide Nitrosamines, A Special Report, 1977. Warsaw.
Jacques, G.L. and Harvey, R.G., 1979. Weed Sci., 27: 450-455.
Jordan, L.S. and Day, B.E., 1962. Weeds, 10: 212-215.
Juo, A.S.R. and Oginni, O.O., 1978. J. Environ. Qual., 7: 9-12.

Karickhoff, S.E. and Brown, D.S., 1978. J. Environ. Qual., 246-252.
Kaufman, D.D., Plimmer, J.R., Kearney, P.C., Blake, J. and Guardra, F.S., 1968. Weed Sci., 16: 266-272.
Kearney, P.C., Sheets, P.J. and Smith, J.W., 1964. Weeds, 12: 83-87.
Kearney, P.C., Oliver, J.E., Helling, C.S., Isensee, A.R. and Kontson, A., 1976. Abst. 172nd Am. Chem. Soc. Natl. Meeting, San Francisco, California.
Kearney, P.C., Oliver, J.E., Helling, C.S., Insensee, A.R. and Kontson, A., 1977. J. Agric. Food Chem., 25: 1177-1180.
Kemp, T.R., Stoltz, L.P., Herron, J.W. and Smith, W.T., 1969. Weed Sci., 17: 444-446.
Kennedy, J.M. and Talbert, R.E., 1977. Weed Sci., 25: 373-381.
Keys, C.H. and Friesen, H.A., 1968. Weed Sci., 16: 341-343.
Khan, S.U., 1973a. Can. J. Soil Sci., 53: 199-204.
Khan, S.U., 1973b. Can. J. Soil Sci., 53: 429-434.
Khan, S.U., 1973c. J. Soil Sci., 24: 244-248.
Khan, S.U., 1973d. J. Environ, Qual., 2: 415-417.
Khan, S.U., 1974a. J. Environ. Qual., 3: 202-206.
Khan, S.U., 1974b. Soil Sci., 118: 339-343.
Khan, S.U., 1974c. Environ. Sci. Technol., 8: 236-238.
Khan, S.U., 1974d. Can. J. Soil Sci., 54: 235-237.
Khan, S.U., 1974e. Residue Rev., 52: 1-26.
Khan, S.U., 1977a. Can. J. Soil Sci., 57: 9-13.
Khan, S.U., 1977b. In: Fate of Pollutants in the Air and Water Environment, Part II, John Wiley & Sons., New York, N.Y., pp. 367-391.
Khan, S.U., 1978. Pestic. Sci., 9: 39-43.
Khan, S.U. and Marriage, P.B., 1979. J. Agric. Food Chem., 27: 1398-1400.
Khan, S.U. and Mazurkewich, R., 1974. Soil Sci., 118: 339-343.
Khan, S.U. and Schnitzer, M., 1971. Can. J. Chem., 13: 2302-2309.
Khan, S.U. and Schnitzer, M., 1972. Geochim. Cosmochim. Acta, 36: 745-754.
Khan, S.U. and Schnitzer, M., 1978. J. Environ. Sci. Health, B13: 299-310.
Khan, S.U. and Young, J.C., 1977. J. Agric. Food Chem., 25: 1430-1432.
Khan, S.U., Belanger, A., Hogue, E.J., Hamilton, H.A. and Mathur, S.P., 1976. Can. J. Soil Sci., 56: 407-412.
Kirk, R.E. and Wilson, M.C., 1960. J. Econ. Entomol., 53: 771-774.
Knight, B.A.G. and Denny, P.J., 1970. Weed Res., 10: 40-48.
Knight, B.A.G. and Tomlinson, T.E., 1967. J. Soil Sci., 18: 233-243.
Kodama, H. and Schnitzer, M., 1971. Can. J. Soil Sci., 51: 509-512.
Konrad, J.G. and Chesters, G., 1969. J. Agric. Food Chem., 17: 226-230.
Konrad, J.G., Armstrong, D.E. and Chesters, G., 1967. Agron. J., 59: 591-594.
Konrad, J.G., Chesters, G. and Armstrong, D.E., 1969. Soil Sci. Soc. Am. Proc., 33: 259-262.
Koren, E.C., Foy, L. and Ashton, F.M., 1968. Weed Sci., 16: 172-175.
Koren, E.C., Foy, L. and Ashton, F.M., 1969. Weed Sci., 17: 148-153.
Kuwahara, M., Kato, N. and Munakata, K., 1965. Agric. Biol. Chem. (Tokyo), 29: 880.
Lambert, S.M., 1967. J. Agric. Food Chem., 15: 572-576.
Lambert, S.M., 1968. J. Agric. Food Chem., 16: 340-343.
Lambert, S.M., Porter, P.E. and Schieferstein, R.H., 1965. Weeds, 13: 185-190.
Lavy, T.L., 1970. Weed Sci., 18: 53-56.

Leenheer, J.A. and Ahlrichs, J.L., 1971. Soil Sci. Soc. Am. Proc., 35: 700-705.
Leistra, M., 1973. Residue Rev., 49: 87-130.
Leopold, A.C., Van Schaik, P. and Neal, M. 1960. Weeds, 8: 48-54.
Letey, J. and Oddson, J.K., 1972. In: C.A.I. Goring and J.W. Hamaker (Editors), Organic Chemicals in the Soil Environment, Vol. 1, Dekker, New York, N.Y., pp. 399-440.
Letey, J. and Farmer, W.J., 1974. In: W.D. Guenzi (Editor), Pesticides in Soil and Water, Soil Sci. Soc. Am. Inc., Publ., Madison, Wisc., pp. 67-97.
Li, G.C. and Felbeck, G.T. Jr., 1972a. Soil Sci. 113: 140-148.
Li, G.C. and Felbeck, G.T. Jr., 1972b. Soil Sci., 114: 201-209.
Liang, T.T. and Lichtenstein, E.P., 1976. J. Agric. Food Chem., 24: 1205-1210.
Lichtenstein, E.P. and Schulz, K.R., 1964. J. Econ. Entomol., 57: 618-627.
Lindstrom, F.T., Boersma, L. and Gardiner, H., 1968. Soil Sci., 106: 107-113.
Lopez-Gonzales, J.D. and Valenzuela-Calahorro, C., 1970. J. Agric. Food Chem., 18: 520-523.
Lowe, L.E., (1978). In: M. Schnitzer and S.U. Khan (Editors), Soil Organic Matter, Elsevier Scientific Publishing Co., New York, N.Y., pp. 65-93.
McGlamery, M.D. and Slife, F.W., 1966. Weeds, 14: 237-239.
Marshall, C.E., 1964. The physical Chemistry and Mineralogy of Soils, Vol. I. John Wiley & Sons, New York, N.Y., 388 pp.
Maslennikova, W.G. and Kruglow, J.W., 1975. Roczniki Gleboznawcze, 26: 25-29.
Massini, P., 1961. Weed Res., 1: 142-146.
Melnikov, N.N., 1971. Chemistry of Pesticides. Springer-Verlag, New York, N.Y., 480 pp.
Menges, R.M., 1964. Weeds, 12: 236-237.
Menn, J.J., McBain, J.B., Adelson, B.J. and Patchett, G.G., 1965. J. Econ. Entomol., 58: 875-878.
Metcalf, R.L., 1971. In: R. White-Stevens (Editor), Pesticides in the Environment, Dekker, New York, N.Y., pp. 1-144.
Miller, R.M. and Faust, S.d., 1972. Environ. Letters, 2: 183-194.
Mills, A.L. and Alexander, M., 1976. J. Environ. Qual., 5: 437-440.
Mingelgrin, U. and Yaron, B., 1974. Proc. Soil Sci. Soc. Am., 38: 914-917.
Mingelgrin, U., Saltzman, S. and Yaron, B. 1977. Soil Sci. Soc. Am. J., 41: 519-523.
Mingelgrin, U., Yariv, S. and Saltzman, S., 1978. Soil Sci. Soc. Am. J., 42: 664-665.
Mingelgrin, U., Kliger, L. and Saltzman, S., 1979. Pestic. Sci., 10: 133-138.
Moilanen, K.W., Crosby, D.G., Soderquist, C.J. and Wong, A.S. 1975. In: R. Haque and V.H. Freed (Editors), Enviromental Dynamics of Pesticides, Plenum Press, New York, N.Y., pp. 45-59.
Morita, H., 1976. Can. J. Soil Sci., 56: 105-109.
Mortland, M.M., 1966. Clay Minerals, 6: 143-156.
Mortland, M.M., 1968. J. Agric. Food Chem., 16: 706-707.
Mortland, M.M., 1970. Adv. Agron., 22: 75-117.
Mortland, M.M. and Meggitt, W.F., 1966. J. Agric. Food Chem., 14: 126-129.
Mortland, M.M. and Raman, K.V., 1967. J. Agric. Food Chem., 15: 163-167.
Nakamoto, K., 1963. Infrared Spectra of Inorganic and Coordination Compounds. John Wiley & Sons, New York, N.Y., 184 p.

Nearpass, D.C., 1965. Weeds, 13: 314-316.
Nearpass, D.C., 1969. Soil Sci. Soc. Am. Proc., 33: 524-528.
Nearpass, D.C., 1971. Soil Sci. Soc. Am. Proc., 35: 64-68.
Nearpass, D.C., 1972. Soil Sci. Soc. Am. Proc., 36: 606-610.
Nearpass, D.C., 1976. Soil Sci., 121: 272-277.
Niemann, P. and Mass, G., 1972. Schriftenr. Ver. Mass. Boden-Lufthyg., Berlin-Dahlem, H37: 155-165.
Nilles, G.P. and Zabik, M., 1975. J. Agric. Food Chem., 23: 410-415.
Nyquist, R.A. and Potts, W.J., 1961. Spectrochim. Acta., 17: 679-697.
O'Connor, G.A. and Anderson, J.U., 1974. Soil Sci. Soc. Am. Proc., 38: 433-436.
Oddson, J.K., Letey, J. and Weeks, L.V., 1970. Soil Sci Soc. Am. Proc., 34: 412-417.
Oliver, J.E. and Kontson, A., 1978. Bull. Environ. Contam. Toxicol., 20: 170-173.
Oliver, J.E., Kearney, P.C., and Kontson, A., 1979. J. Agric. Food Chem., 27: 887-891.
O'Toole, M.A., 1966. Irish Crop Protec. Conf. Proc., pp. 35-39.
Pancholy, S.K., 1978. Soil Biol. Biochem., 10: 27-32.
Parochetti, J.V. and Warren, G.F., 1966. Weeds, 14: 281-285.
Parochetti, J.V. and Dec, G.W. Jr., 1978. Weed Sci., 26: 153-156.
Paschal, D.C. and Neville, M.E., 1976. J. Environ. Qual., 5: 441-443.
Pick, M.E., 1973. The Interaction of Bipyridylium Salts and Clay Minerals. Ph.D. Thesis, Univ. Birmingham, U.K.
Pierce, Jr., R.H., Olney, C.E. and Felbeck, Jr. G.T., 1971. Environ. Letters, 1: 157-172.
Pierce, Jr., R.H., Olney, C.E. and Felbeck, Jr. G.T., 1974. Geochim. Cosmochim. Acta, 38: 1061-1073.
Plimmer, J.R., 1970. Residue Rev., 33: 47-74.
Plimmer, J.R., Kearney, P.C., Kaufman, D.D. and Guardia, F.S., 1967. J. Agric. Food Chem., 15: 996-999.
Plimmer, J.R., Kearney, P.C., Chisaka, H., Young, J.B. and Klingebiel, U.I., 1970. J. Agric. Food Chem., 18: 859-861.
Prost, R., Gerstl, Z., Yaron, B. and Chaussidon, J. 1976. In: Behavior of Pesticides in Soil, ARO-Volcani Center, Bet Dagan, Israel, pp. 27-33.
Rake, D.W., 1961. Weed Soc. Am. Abst., 88.
Rao, Y.R. and Sethunathan, N. 1979. J. Environ. Sci. Health, B14: 335-351.
Rhodes, R.C., Belasco, I.J. and Pease, H.L., 1970. J. Agric. Food Chem., 18: 524-528.
Riekerk, H. and Gessel, S.P., 1968. Soil Sci. Soc. Am. Proc., 32: 595-600.
Riley, D., Wilkinson, W. and Tucker, B.V., 1976. In: D.D. Kaufman, G.G. Still, G.D. Paulson and S.K. Bandal (Editors), Bound and Conjugated Pesticide Residues, ACS Symp. Ser., 29, pp. 301-353.
Ross, R.D. and Crosby, D.G., 1973. J. Agric. Food Chem., 21: 335-337.
Ross, R.D., Morrison, J., Rounbehler, D.P., Fan, S. and Fine, D.H., 1977. J. Agric. Food Chem., 25: 1416-1418.
Russell, J.D., Cruz, M. and White, J.L., 1968a. J. Agric. Food Chem., 16: 21-24.
Russell, J.D., Cruz, M., White, J.L., Bailey, G.W., Payne, W.R., Jr., Pope, J.D., Jr. and Teasley, J.I., 1968b. Science, 160: 1340-1342.
Saltzman, S. and Kliger, L., 1979. J. Environ. Sci. Health, B14: 353-366.
Saltzman, S. and Yariv, S., 1976. Soil Sci. Soc. Am. J., 40: 34-38.

Saltzman, S., Mingelgrin, U. and Yaron, B., 1976. J. Agric. Food Chem., 24: 739-743.
Saltzman, S., Kliger, L. and Yaron, B., 1972. J. Agric. Food Chem., 20: 1224-1226.
Saltzman, S., Yaron, B. and Mingelgrin, U., 1974. Soil Sci. Soc. Am. Proc., 38: 231-234.
Saunders, D.G., Mosier, J.W., Gray, J.E. and Loh, A., 1979. J. Agric. Food Chem., 27: 584-589.
Savage, K.E. and Wanchope, R.D., 1974. Weed Sci., 22: 106-110.
Schliebe, K.A., Burnside, O.C. and Lavy, T.L., 1965. Weeds, 13: 321-325.
Schnitzer, M. and Khan, S.U., 1972. Humic Substances in the Environment, Dekker, New York, N.Y., 327 pp.
Schnitzer, M. and Khan, S.U., 1978. Soil Organic Matter, Elsevier Scientific Publication Company, Amsterdam, 319 pp.
Schwartz, H.G., Jr., 1967. Environ. Sci. Technology, 1: 332-337.
Scott, D.C. and Weber, J.B., 1967. Soil Sci., 104: 151-518.
Scott, H.D. and Phillips, R.E., 1972. Soil Sci. Soc. Am. Proc., 36: 714-719.
Shearer, R.C., Letey, J., Farmer, W.J. and Klute, A., 1973. Soil Sci. Soc. Am. Proc., 37: 189-193.
Sheldon, D.W. and Day. E.W. Jr., 1979. J. Agric. Food Chem., 27: 1075-1080.
Shin, Y.O., Chodan, J.J. and Wolcott, A.R., 1970. J. Agric. Food Chem., 18: 1129-1133.
Singhal, J.P. and Bansal, V., 1978. Soil Sci., 360-363.
Singhal, J.P. and Singh, C.P., 1976. J. Agric. Food Chem., 24: 307-310.
Skipper, H.D., Volk, V.V. and Frech, R., 1976. J. Agric. Food Chem., 24: 126-129.
Skipper, H.D., Volk, V.V., Mortland, M.M. and Rainan, K.V., 1978. Weed Sci., 26: 46-50.
Smith, C.A., Iwata, Y. and Gunther, F.A., 1978. J. Agric. Food Chem., 26: 959-962.
Spencer, W.F., 1970. In: pesticides in the Soil: Ecology, Degradation and Movement, Mich. State Univ., E. Lansing, Mich., pp. 120-128.
Spencer, W.F. and Cliath, M.M., 1969. Environ. Sci. Technol., 3: 670-674.
Spencer, W.F. and Cliath, M.M., 1970a. J. Agric. Food Chem., 18: 529-530.
Spencer, W.F. and Cliath, M.M., 1970b. Soil Sci. Soc. Am. Proc., 34: 574-578.
Spencer, W.F. and Cliath, M.M., 1972. J. Agric. Food Chem., 20: 645-649.
Spencer, W.F., Cliath, M.M. and Farmer, W.J., 1969. Soil Sci. Soc. Am. Proc., 33: 509-511.
Spencer, W.F., Farmer, W.J. and Cliath, M.M., 1973. Residue Rev., 49: 1-47.
Spencer, W.F., Cliath, M.M., Farmer, W.J. and Shepherd, R.A., 1974. J. Environ. Qual., 3: 126-129.
Stevenson, F.J., 1966. J. Am. Oil Chem. Soc., 43: 203-210.
Stevenson, F.J., 1972. J. Environ. Qual., 1: 333-343.
Stevenson, F.J., 1976. In: D.D. Kaufman, G.G. Still, G.D. Paulson and S.K. Bandal (Editors), Bound and Conjugated Pesticides Residues, ACS Smyp. Ser., 29, pp. 180-207.
Su, Y.H. and Lin, H.C., 1971. Chem. Abstr., 74: 301.
Sullivan, Jr., J.D. and Felbeck, Jr., G.T., 1968. Soil Sci., 106: 42-52.

Supak, J.R., Swoboda, A.R. and Dixon, J.B. 1978. Soil Sci. Soc. Am. J., 42: 244-248.
Swoboda, A.R. and Thomas, G.W., 1968. J. Agric. Food Chem. 16: 923-927.
Tabatabai, M.A., 1974. Commun. Soil Sci. Plant Anal., 5: 569-578.
Tate, R.L. and Alexander, M., 1974. Soil Sci., 118: 317-321.
Tate, R.L. and Alexander, M., 1975. J. Natl. Cancer Inst., 54: 327-330.
Tate, R.L. and Alexander, M., 1976. J. Environ. Qual., 5: 131-133.
Theng, B.K.G., 1974. The Chemistry of Clay Organic Reactions. A. Holger, London, 260 pp.
Thompson, J.M., 1968. Ph.D. Thesis, University of Birmingham, England.
Toth, S.J., 1950. In: F.E. Bear (Editor), Chemistry of the Soil, Reinhold Publishing Corporation, New York, N.Y., pp. 85-106.
Trichell, D.W., Morton, H.L. and Merkle, M.G., 1968. Weed Sci., 16: 447-449.
Tsunoda, H., 1965. J. Sci. Soil Manure, Jap., 36: 177-181.
Tucker, B.V., Pack, D.E. and Ospenson, J.N., 1967. J. Agric. Food Chem., 15: 1005-1008.
Tucker, B.V., Pack, D.E. Ospenson, J.N. Omid, A. and Thomas, W.D., Jr., 1969. Weed Sci., 17: 448-451.
Upchurch, R.P. and Mason, D.D., 1962. Weeds, 10: 9-14.
Van Genuchten, M. Th., Davidson, J.M. and Wierenga, P.J., 1974. Soil Sci. Soc. Am. Proc., 38: 29-35.
Van Olphen, H., 1963. An Introduction to Clay Colloid Chemistry. Interscience Publishers, New York, N.Y., 301 pp.
Viltos, A.J., 1952. Proc. Northeast Weed Contr. Conf., 6: 57-62.
Viltos, A.J., 1953. Contrib. Boyce Thompson Inst., 17: 127-149.
Wahid, P.A. and Sethunathan, N., 1978. J. Agric. Food Chem., 26: 101-105.
Walker, A. and Crawford, D.V., 1968. In: Isotopes and Radiation in Soil Organic-Matter Studies. I.A.E.A., Vienna, pp. 91-105.
Walker, A. and Crawford, D.V., 1970. Weed Res., 10: 126-132.
Wang, W.G., 1968. Diss. Abstr., B29(3): 904B-905B.
Ward, T.M. and Upchurch, R.P., 1965. J. Agric. Food Chem., 13: 334-340.
Warshaw, R.L., Burcar, P.J. and Goldberg, M.C., 1969. Environ. Sci. Technol., 3: 271-273.
Weber, J.B., 1966. Am. Miner., 51: 1657-1661.
Weber, J.B., 1970a. Residue Rev., 32: 93-130.
Weber, J.B., 1970b. Soil Sci. Soc. Am. Proc., 34: 401-404.
Weber, J.B., 1972. In: R.F. Gould (Editor), Fate of Organic Pesticides in the Aquatic Environment, Am. Chem. Soc., 111: pp. 55-120.
Weber, J.B., Ward, T.M. and Weed, S.B., 1968. Soil Sci. Soc. Am. Proc., 32: 197-200.
Weber, J.B., Perry, P.W. and Upchurch, R.P., 1965. Soil Sci. Soc. Am. Proc., 29: 678-688.
Weber, J.B., Weed, S.B. and Ward, T.M., 1969. Weed Sci., 17: 417-421.
Weber, J.B., Weed, S.B. and Waldrep, T.W., 1974. Weed Sci., 22: 454-459.
Weber, W.J. and Gould, J.P., 1966. Adv. Chem. Ser., 60: 280-304.
Weed, S.B. and Weber, J.B., 1968. Am. Miner., 53: 478-490.
Weed, S.B. and Weber, J.B., 1969. Soil Sci. Soc. Am. Proc., 33: 379-382.
Weil, L., Durê, G. and Quentin, K.E., 1973. Z. Wasser Abwasser Forsch., 6: 107-112.

White, A.W., Barnett, A.P., Wright, B.G. and Holladay, J., 1967. Environ. Sci. Tech., 1: 740-744.
Willis, G.H., Parr, J.F. and Smith, S., 1971. Pestic. Monit. J., 4: 204-208.
Wolf, D.C., Johnson, R.S., Hill, G.D., and Varner., 1958. Herbicidal Properties of Neburon. Proc. N.C. Weed Control Conf., 15: 7.
Wright, W.L. and Warren, G.F., 1965. Weeds, 13: 329-331.
Yaron, B., 1975. Soil Sci. Soc. Am. Proc., 39: 639-643.
Yaron, B., 1978. Soil Sci. 125: 210-216.
Yaron, B. and Saltzman, S., 1978. Residue Rev., 69: 1-34.
Yih, R.Y., Swithenbank, C. and MacRae, D.H., 1970. Weed Sci., 18: 604-607.
Young, J.C. and Khan, S.U., 1978. J. Environ. Sci. Health, B13: 59-72.
Young, J.C., Khan, S.U.and Marriage, P.B., 1977. J. Agric. Food Chem., 25: 918-922.
Yuen, Q.H. and Hilton, H.W., 1962. J. Agric. Food Chem., 10: 386-392.

Chapter 4

MICROBIAL PROCESSES AFFECTING PESTICIDES IN SOIL

Microbial degradation plays an important role in affecting the fate and behavior of many pesticides in soil. Factors affecting the microbial degradation of pesticides in soil include pH, time, temperature, adsorption, moisture and soil type. Degradation of pesticides has been followed in the soil by several methods, such as extraction and chemical analysis, bioassays, oxygen uptake, and evolution of carbon dioxide. Some of these methods have been used in studies comparing sterile vs. nonsterile soils.

Several reviews have been published dealing with specific structural characteristics of pesticides that are associated with or that prevent microbial decomposition (Alexander and Aleem, 1961; Alexander, 1965; Kaufman and Plimmer, 1972). Kaufman and Plimmer (1972) discussed the structure-activity-degradability interrelations of several major pesticide classes. Kaufman (1974) reviewed the degradation of pesticides by soil microorganisms. Several other reviews on the metabolism of pesticides by microorganisms have also been published elsewhere (Kaufman and Kearney, 1970; Matsumura and Boush, 1971; Laveglia and Dahm, 1977).

This chapter examines the processes involved in the microbial degradation of pesticides.

4.1. HERBICIDES

Considerable information is available on the microbial metabolism of herbicides in soils. Some of the important herbicides that have received major attention in microbial metabolism are discussed below.

4.1.1. Arsenicals

The two important organic arsenical herbicides, MSMA and cacodylic acid, are metabolized in the soil by microbiological activity. Von Endt et al. (1968) observed that MSMA-^{14}C was oxidized slowly to $^{14}CO_2$ in Hagerstown silty clay loam. They concluded that soil microorganisms played some role in the decomposition process. Several actinomycetes, a fungus and several bacteria were isolated using soil enrichment techniques.

Cacodylic acid degradation is caused by two mechanisms: cleavage of the C-As bond(s) and reduction to a volatile organoarsenical, probably dimethylarsine or an oxide (Woolson and Kearney, 1973). The degradation is slow, with 15 to 80% of the ^{14}C activity lost in 32 weeks, depending on the soil type.

4.1.2. Organophosphates

Complete and rapid microbiological degradation of glyphosate occurs in soils and the only significant metabolite, aminomethylphosphoric acid, also undergoes rapid degradation (Rueppel et al., 1977). The microbial degradation of glyphosate in soil may be stimulated by adding phosphate, or reduced by adding Fe^{3+} and Al^{3+} (Moshier and Penner, 1978). A thin layer chromatographic method for the separation of metabolites, aminomethylphosphoric acid, glycine, and sarcosine has been described by Sprankel et al. (1978).

4.1.3. Phenoxys

The metabolism of phenoxyalkanoic acids by soil mircroorganisms has been the subject of several extensive reviews (Kaufman, 1970; Helling et al., 1971; Loos, 1975). The organisms that metabolize various chlorinated members of this herbicide family include species of *Pseudomonas*, *Achromobacter*, *Flavobacterium*, *Corynebacterium*, *Arthrobacter*, and *Sporocytophaga* (Loos, 1975). The major metabolic reactions associated with phenoxyalkanoic acids include: (1) ring hydroxylation, (2) cleavage of the ether linkage; (3) ring cleavage; (4) dehalogenation, and (5) β-oxidation of the long chain aliphatic acid moiety.

The type and position of the ring substituents, and the specific microorganism involved in degradation will influence the position of ring hydroxylation in phenoxyalkanoates. Ring hydroxylation by *Aspergillus niger* of omega-substituted, nonchlorinated phenoxyalkanoates occurs in the *ortho* and *para* ring position (Byrde and Woodcock, 1957). *Aspergillus niger* hydroxylates 2,4-D (9) and MCPA (10) to 5-hydroxy-2,4-D (159) and 5-hydroxy-MCPA (166), respectively (Faulkner and Woodcock, 1961) (*Scheme 4.1*). However, the site of hydroxylation may be different with *Pseudomonas* since 6-hydroxy-2,4-D and 6-hydroxy-MCPA are produced from 2,4-D and MCPA, respectively (Loos, 1975).

Helling et al. (1968) demonstrated that in phenoxyalkanoic acids the cleavage of ether linkage occurs between the aliphatic side chain and ether-oxygen atom. Phenoxy-^{18}O-acetate was metabolized to ^{18}O-phenol by cell free extracts of an *Arthrobacter* sp. in an O_2 requiring process. Enzymatic cleavage of 2,4-D to 2,4-dichlorophenol (160) involves oxidation of the methylene carbon and formation of the α-hydroxy-2,4-D derivative, which is then cleaved to 2,4-dichlorophenol (160) and glyoxylate (Tiedje and Alexander, 1969).

The ring structure of several ω-phenoxyalkanoic acids is lost during their degradation in soil (Alexander and Aleem, 1961). Ring cleavage proceeds through the intermediate formation of the corresponding catechols from the phenols with subsequent ring opening and formation of a muconic acid (Kaufman, 1974). Production of 3,5-dichlorocatechol (161) from 2,4-dichlorophenol (160) by enzymatic processes requires both O_2 and NADPH (Bollage et al., 1968). 2,4-D, MCPA, and 4-chlorophenoxyacetic acid yield their corresponding chloromuconic acids (Fernley and Evans, 1959; Gaunt and Evans, 1961). Enzyme preparations from *Arthrobacter* sp. catalyze the conversion of 3,5-dichlorocatechol (161) to 2,4-dichloromuconic acid (162) (Tiedje et al., 1969). The final path to CO_2 lies through chloromaleylacetic acid (163) as an intermediate (Tiedje et al., 1969), which *Arthrobacter* sp. enzymes degrade through maleylacetic acid (164) to succinic acid (165) (Duxbury et al., 1970). By a similar initial cleavage of the phenyl ether linkage, several bacteria have been reported to break MCPA (10) to 2-methyl-4-chlorophenol (166) (Bollag et al., 1967), and the corresponding catechol has been found in *Pseudomonas* (Gaunt and Evans, 1971).

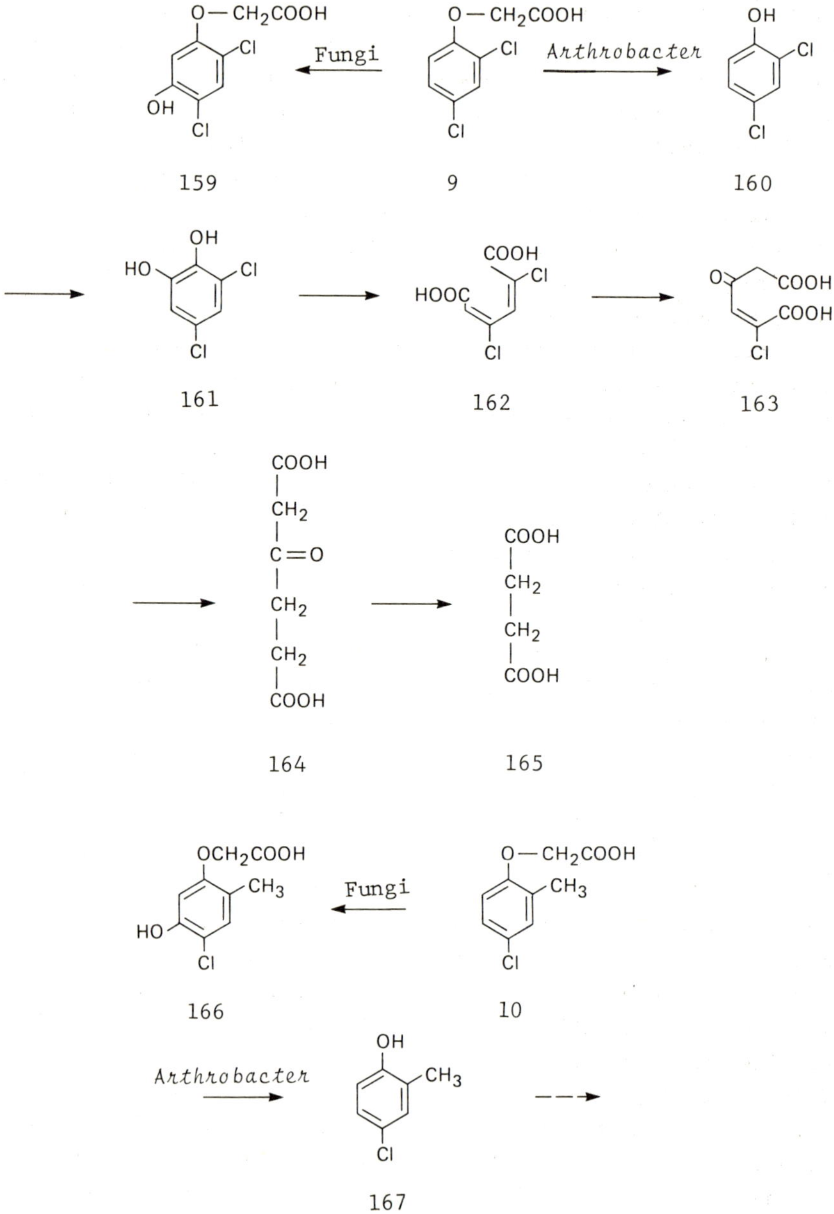
Fungi
Arthrobacter
159
9
160
161
162
163
164
165
Fungi
166
10
Arthrobacter
167

Scheme 4.1

β-Oxidation results in the formation of phenoxyalkanoic acids of shorter chain lengths by a series of well studied reactions. Gutenmann et al. (1964) showed that β-oxidation of 2,4-dichlorophenoxyalkanoic acids occurred in a natural soil. Gutenmann and Lisk (1964) subsequently obtained evidence that the β-oxidation of 2,4-DB in soil proceeded via the expected first intermediate, 4-(2,4-dichlorophenoxy) crotonic acid. Factors that hinder β-oxidation in plants also inhibit β-oxidation in microorganisms (Loos, 1975).

Smith and Phillips (1976) demonstrated that the initial step in 2,4-DB metabolism by *Phytophthora megasperma* did not include β-oxidation of 2,4-DB.

4.1.4. Benzoic Acids

Dewey et al. (1962) observed biodegradation of TBA (12) in nonsterile soil with release of inorganic chloride. Horvath (1971) suggested that *Brevibacterium* sp. degraded TBA by a process involving ring hydroxylation in the 4-position followed by decarboxylation and dehalogenation to yield 3,5-dichlorocatechol (161) as a toxic end product.

Numerous studies have indicated that chloramben is subject to microbial degradation in soil (MacRae and Alexander, 1964; Corbins and Upchurch, 1967; Sheets et al., 1968). Soil microflora break down chloramben (Rauser and Switzer, 1962) and the carboxyl group is removed slowly but steadily (MacRae and Alexander, 1965; Wildung et al., 1968).

Microbial degradation in the soil is also considered an important route of dissipation for other benzoic acid herbicides such as dicamba (Wurzer and Corbins, 1968), dichlobenil (Smith and Sheets, 1967) and chlorthiamid (Beynon and Wright, 1968).

4.1.5. Pyridine Acids

Picloram (16) the only prominent member of the pyridine acids family, is degraded in soil by microorganisms (Meikle et al., 1966; Youngson et al., 1967; Grover, 1967). The only metabolite detected in soil is 6-hydroxypicloram (168) (Youngson et al., 1967) (*Scheme 4.2*). Decarboxylation as a possible degradation reaction

16 168

Scheme 4.2

is indicated by the detection of small amounts of $^{14}CO_2$ evolved from carboxyl ^{14}C-labeled picloram treated soil.

4.1.6. Amides

Soil fungi *Trichoderma viride* and *Aspergillus candidus* degrade diphenamide resulting in the formation of *N*-methyl-2,2-diphenylacetamide and 2,2-diphenylacetamide (Kesner and Ries, 1967). Several isolated soil microorganisms dehalogenate CDAA (Kaufman and Blake, 1973). Degradation of alachlor in a sandy loam soil results in the removal of methoxymethyl substituent from the amide nitrogen (Hargrove and Merkle, 1971).

Soil incubation studies have shown that alachlor (21) is biodegraded relatively rapidly in soils (Beestman and Deming, 1974). Kaufman and Blake (1973) found that *Fusarium oxysporum* released some chloride from alachlor but did not produce aniline intermediates. Taylor (1972) observed that the common soil fungus *Chaetomium globosum* rapidly metabolized ring labeled (^{14}C)-alachlor without producing $^{14}CO_2$. Degradation of alachlor by *Chaetomium globosum* produces metabolites 2,6-diethyl-*N*-(methoxymethyl) aniline, 2,6-diethylaniline, 1-chloroacetyl-2,3-dihydro-7-ethylindole and 2-chloro-2,6-diethylacetanilide. Incubation of *C. globosum* with 2-chloro-2,6-diethylacetanilide, 2,6-diethylaniline, and monochloroacetic acid results in further degradation of these products (Tiedje and Hagedron, 1975).

4.1.7. Thiocarbomates, Phenylcarbamates and Acylanilides

Soil microorganisms metabolize thiocarbamate herbicides when incorporated in soil (Sheets, 1959; MacRae and Alexander, 1965;

Kaufman, 1967). In the microbial degradation of thiocarbamate and dithiocarbamate herbicides, several sites of attack are possible, e.g. the alkyl groups, the amide linkage, or the ester linkage (Fang, 1975). The thiocarbamate molecule is probably hydrolyzed at the ester linkage with the formation of mercaptan, CO_2, and amine (Kaufman, 1967) (*Scheme 4.3*). The mercaptan could then be converted to alcohol and further oxidized. ^{14}C-Labeled EPTC applied to soil is metabolized by soil microorganisms, however, the rate of $^{14}CO_2$ release from the ethyl moiety of the molecule is slow in comparision to the rate of inactivation (MacRae and Alexander, 1965).

The phenylcarbamate herbicide, chlorpropham is degraded by soil bacteria identified as species of *Pseudomonas*, *Flavobacterium*, *Agrobacterium* and *Achromobacter* (Kaufman and Kearney, 1965).

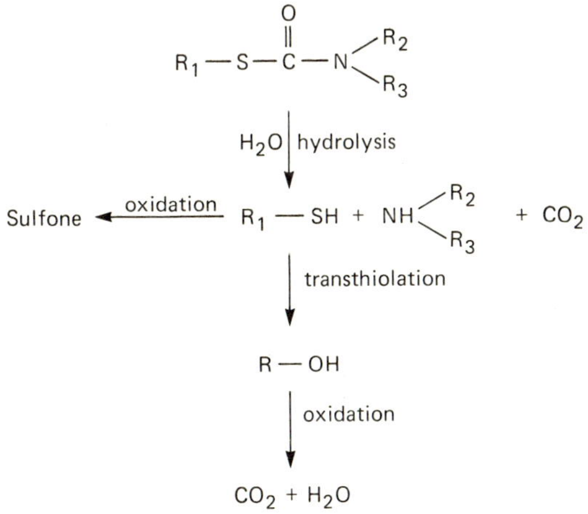

Scheme 4.3

Kearney (1965) observed that an enzyme isolated from *Pseudomonas striata* cleaved the carbamate linkage to yield 3-chloroaniline, CO_2 and isopropanol.

Penicillium piscarium converts propanil (19) to 3,4-dichloroaniline (169) and *Geotrichum candidum* converts 169 to 171 (Bordeleau and Bartha, 1971). Chisaka and Kearney (1970) and Plimmer et al. (1970b) examined the metabolism of propanil (19) in soils. They confirmed the cleavage of the anilide to 3,4-dichloroaniline (169) and the oxidative catabolism of the propionic acid (170) moiety (*Scheme 4.4*). Further metabolism of the 3,4-dichloroaniline moiety in soils resulted in the formation of 3,3,4,4-tetrachloroazobenzene (171) and a new metabolite identified as 1,3-bis (3,4-dichlorophenyl) triazine (172). Kearney et al. (1970) detected low concentrations of 171 in rice producing soils. The high molecular weight metabolite (172) was isolated from a Japanese soil incubated with propanil (Plimmer et al., 1970b).

4.1.8. Dinitroanilines

Both aerobic and anaerobic metabolic degradation pathways have been proposed for several dinitroaniline herbicides (Probst et al.,

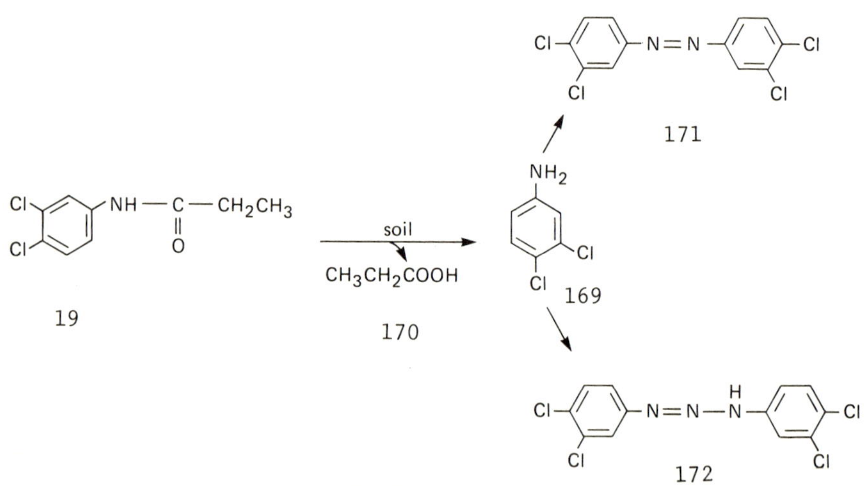

Scheme 4.4

1975). Under aerobic conditions, dealkylation is the first step in the metabolism of trifluralin (35) in soil. Sequential removal of the second alkyl group would yield the dealkylated product. Reduction of the two nitro groups eventually leads to the formation of 3,4,5-triamino-α,α,α-trifluorotoluene (174) (*Scheme 4.5*). Under anaerobic conditions the nitro groups would first be reduced, followed by dealkylation, with the formation again of 3,4,5-triamino-α,α,α-trifluorotoluene (174).

Trifluralin (35) incubated aerobically in soil undergoes dealkylation to monodealkylated (175) and subsequently to the didealkylated product (176) (Wheeler et al., 1979). Furthermore, two benzimidazole derivatives (177) and (178) are also formed (*Scheme 4.6*). Golab et al. (1979) investigated the degradation of trifluralin in field soil over a three year period. Twenty-eight transformation products were isolated and identified. However, none of the isolated transformation products exceeded 3% of the initially applied trifluralin. It has been suggested that a biological break down accounts for only a small fraction of trifluralin degradation (Messersmith et al. 1971).

Soil fungi *Aspergillus fumigatus* Fres. and *Paecilomyces* sp. degrade dinitramine by at least two metabolic pathways (Laanio et al., 1973). Dinitramine (38) is degraded into the corresponding mono- and didealkylated derivatives, but at the same time they can cyclize it to a benzimidazole derivative, 6-amino-2-methyl-7-nitro-5-trifluoromethyl-benzimidazole (180) (*Scheme 4.7*).

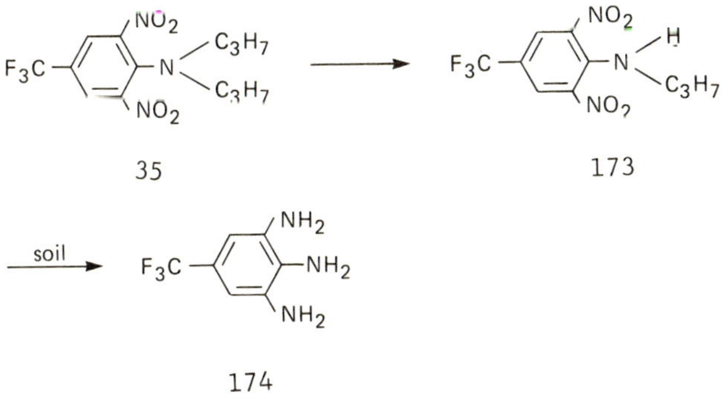

Scheme 4.5

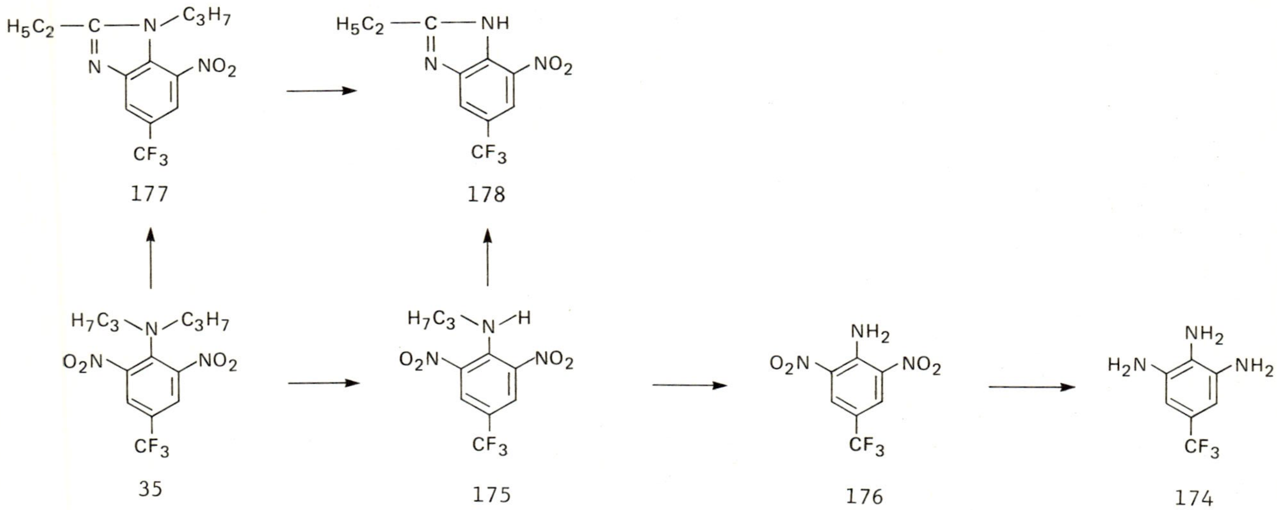

Scheme 4.6

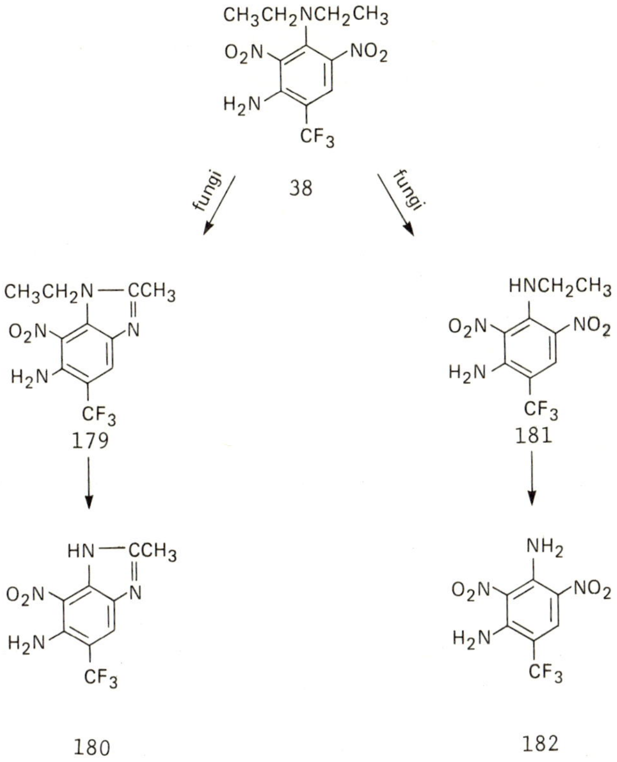

Scheme 4.7

Helling (1976) suggested that as a group, the dinitroaniline herbicides are degraded more rapidly in aerobic than in anaerobic soil.

4.1.9. Bipyridyliums

In soil enrichment cultures of unidentified bacterium, paraquat is first demethylated, followed by ring cleavage, to yield the carboxylated *N*-methylpyridinium ion (Funderburk and Bozarth, 1967). The latter produces methylamine with washed suspensions of a

Achromobacter sp., and after decarboxylation to CO_2 the remaining five carbon atoms give rise to succinate and formate (Wright and Cain, 1972). Baldwin et al. (1966) isolated several microorganisms such as *Corynebacterium fascians* and *Clostridium pasteurianum*, which metabolized paraquat, and specifically a yeast, *Lipomyces starkeyi*, which utilizes paraquat as a sole source of nitrogen.

4.1.10. Uracils

Many uracil herbicides are biodegradable. Soil diphtheroids and *Pseudomonas* sp. present in a wide variety of agricultural soils attack isocil (Reid, 1963). A soil isolate of *Penicillium paraherquei. Abe.* was particularly active in the degradation of bromacil (Torgeson and Mee, 1967).

4.1.11. *s*-Triazines

The metabolism of *s*-triazines by soil microorganisms has been the subject of several reviews (Harris et al., 1968; Kaufman and Kearney, 1970; Esser et al., 1975).

Although *N*-dealkylation appears to be a major pathway for the chloro-*s*-triazines by soil fungi, little information is available on the conversion of *s*-triazines to their hydroxy compounds by microorganisms. Couch et al. (1965) observed that *Fusarium roseum* (LK, Snyder and Hansen) hydrolyzed atrazine to its corresponding hydroxy analogue. Very little is known about the degradation of 2-hydroxy-*s*-triazine by other soil microorganisms. Some of the possible metabolites of atrazine (52) are the hydroxylated analogue of 52, namely hydroxyatrazine (155), partially *N*-dealkylated intermediates of 52 and 155, including 183, 184, 156 and 157, further *N*-dealkylation of which will lead to the formation of compounds 185 and 158. Both of these compounds may then undergo side chain modification, deamination or ring cleavage resulting in the liberation of CO_2 and NH_3 (*Scheme 4.8*).

Several fungi including *Aspergillus fumigatus*, *A. flavipes*, *A. Ustus*, *Rhizopus stolonifer*, *Fusarium moniliforme*, *F. roseum*, *F. oxysporum*, *Penicillium decumbens*, *P. janthinellum*, *P. rugulosum*, *P. luteum* and *Trichoderma viride* degrade atrazine (Kaufman and Blake, 1970). The soil fungus *Aspergillus fumigatus* Fres. metabolizes only the ^{14}C-ethyl groups of simazine while the ring

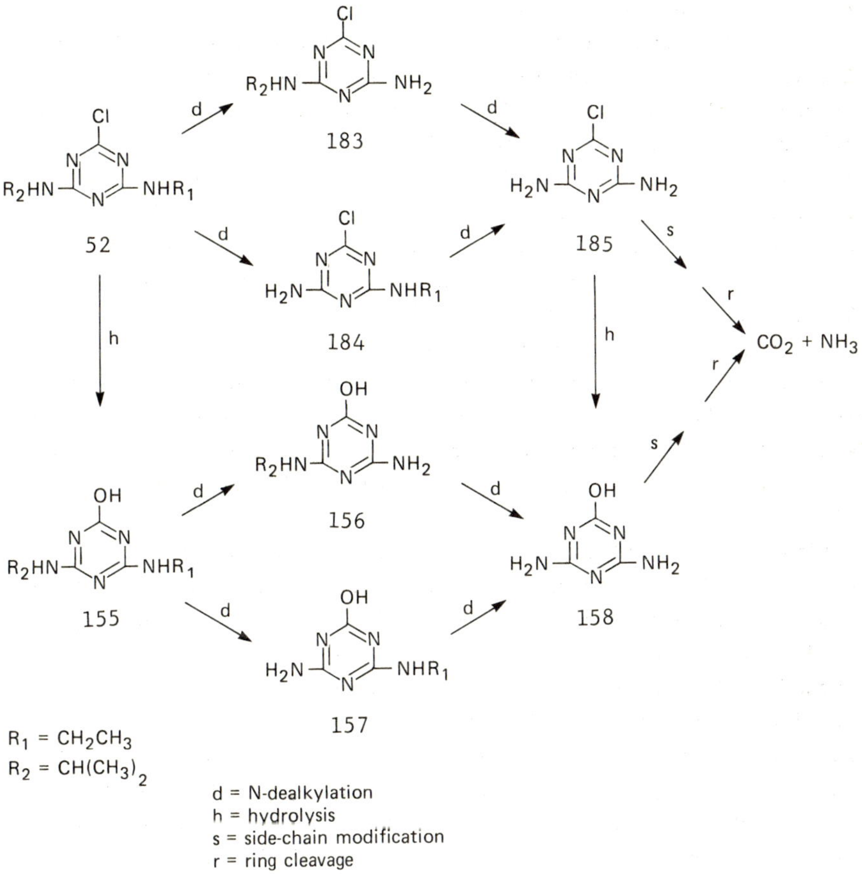

Scheme 4.8

portion remains intact (Kaufman et al., 1963, 1965). Metabolism proceeds by dealkylation, and no ring cleavage occurs. The two degradation products isolated were 2-chloro-4-amino-6-isopropylamino-*s*-triazine (183) and 2-chloro-4-ethylamino-6-amino-*s*-triazine (184) (Kearney et al., 1965). Several representatives of chloro-, methoxy-, and methylthio-*s*-triazine groups also undergo dealkylation with *Aspergillus fumigatus* Fres. (Kaufman and Plimmer, 1971). The ease of alkyl group cleavage by microorganisms decreases in the sequence ethyl, isopropyl, and larger or more branched ethyl groups (Esser et al., 1975).

Khan and Marriage (1977) investigated the metabolism of atrazine (52) in an orchard soil after nine consecutive annual applications of the herbicide. They observed the residues of atrazine (52) and metabolites 155, 156, 157 and 183 in soils even though the samples were taken 2 and 3½ years after the last application of the herbicide. Partial *N*-dealkylation and hydrolysis reactions are involved in the metabolism of atrazine in soil. The existence of compounds 158 and 185 has been reported in soil (Esser et al., 1975). However, degradation of these metabolites in soils occurs very rapidly (Wolf and Martin, 1975).

Hydroxyatrazine (155) was found to be the predominant *s*-triazine residue in the field soil during the spring and autumn (Muir and Baker, 1978). *N*-Deethylated atrazine (183) was also observed as a major metabolite and persisted at relatively high levels in soils.

Plimmer et al. (1970a) examined the degradation of ring- and methylthio-^{14}C labeled prometryn (54) in a silty clay loam maintained under aerobic and flooded conditions. After 6 months, 77% and 86% of the initial ring-^{14}C were present in the aerobic and flooded soils, respectively. Furthermore, 33% and 60% of the methylthio-^{14}C remained after this period. Prometryn sulfoxide, (186), prometryn sulfone (187), and hydroxyprometryn (189) were identified as degradation products (*Scheme 4.9*). Murray and Rieck (1968) observed that bioassay of microbial cultures treated with prometryn also metabolized the herbicide by *Aspergillus niger*, *A. tamaru*, *A. Flavus*, and *A. oryzae*. Prometryn (54) exposed to pure culture of *Aspergillus fumigatus* degrades primarily by *N*-dealkylation to the product 188 (Kaufman and Plimmer, 1971).

4.1.12. Phenylureas

The role of microorganisms in the biodegradation of phenylureas is well established (Geissbuhler et al., 1975). A large number of fungi and bacteria are able to demethylate linuron, monolinuron, diuron and monuron (Schroeder, 1970). *Aspergillus nidulans* is

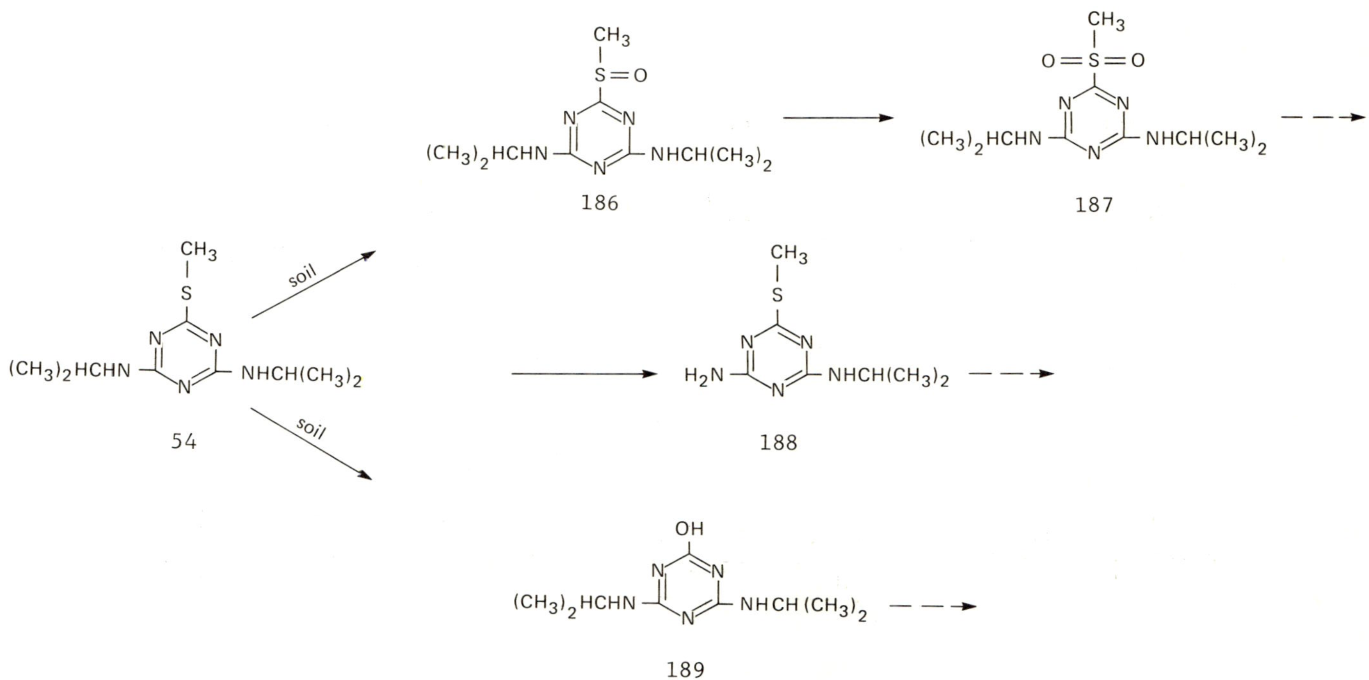
CH3
S=O
(CH3)2HCHN
NHCH(CH3)2
186
CH3
O=S=O
(CH3)2HCHN
NHCH(CH3)2
187
CH3
S
(CH3)2HCHN
NHCH(CH3)2
54
soil
soil
CH3
S
H2N
NHCH(CH3)2
188
OH
(CH3)2HCHN
NHCH(CH3)2
189

Scheme 4.9

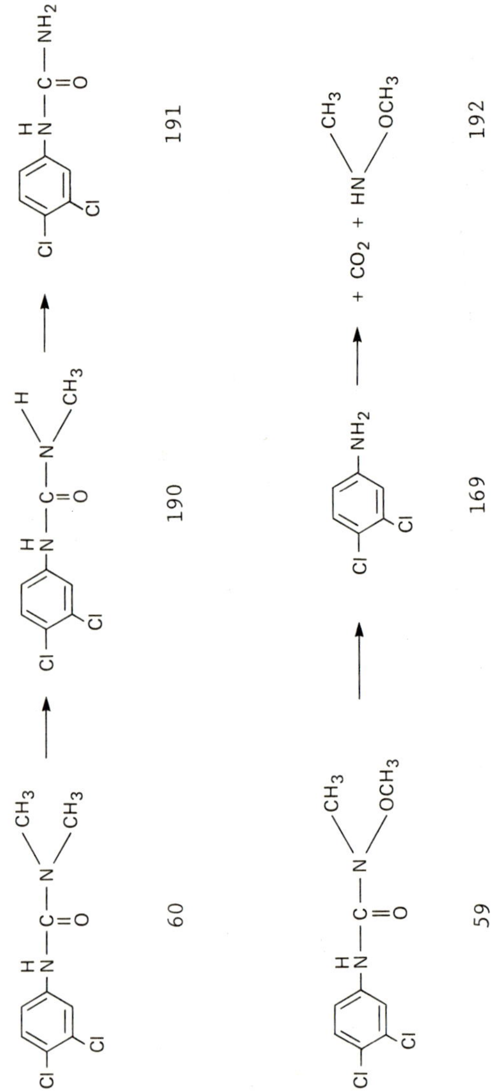

60 190 191

59 169 192

Scheme 4.10

one of the most effective isolates and decomposes more than 50% of the linuron in culture solutions. Metabolism proceeds by successive removal of methyl groups followed by hydrolysis of phenylurea to the corresponding aniline (Geissbühler et al., 1963) (*Scheme 4.10*). Soil degradation of diuron (60) results in the formation of metabolites, 1-methyl-3-(3,4-dichlorophenyl) urea (190) and 3-(3,4-dichlorophenyl) urea (191) (Dalton et al., 1966). Wallnofer (1969) reported that the *N*-methoxy group is more labile in culture solutions of *Bacillus sphaericus* than the *N*-methyl group. *B. sphaericus* metabolized monolinuron, linuron, and metobromuron within a short time, whereas monuron, diuron, fluometuron and methabenzthiazuron appeared to be resistant to decomposition. Engelhardt et al. (1972) observed that *Bacillus sphaericus* isolated from soil produced 3,4-dichloroaniline (169) by hydrolyzing linuron (59) at the amide bond, the side chain yielding CO_2 and *O*,*N*-dimethylhydroxylamine (192).

Viswanathan et al. (1978) carried out long term studies on the fate of 3,4-dichloroaniline (169) in a plant soil system under outdoor conditions. It was observed that the major conversion products formed under laboratory conditions were also formed under outdoor conditions. An acetylation pathway was suggested as evidenced by the formation of 3,4-dichloroacetanilide (193) and 6-hydroxy-3,4-dichloroacetanilide (194). In addition, 3,4,3,4-tetrachloroazobenzene, a metabolite of 3,4-dichloroaniline (195) was also reported in earlier studies (Kearney et al., 1969;

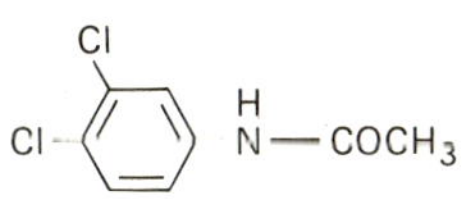

193

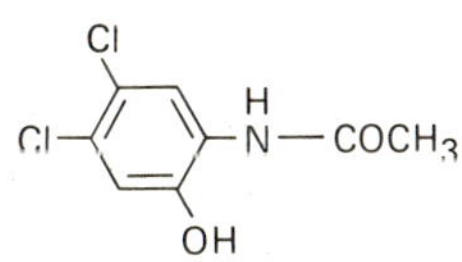

194

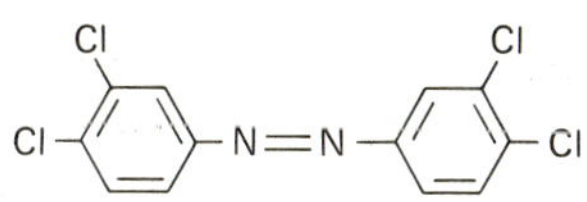

195

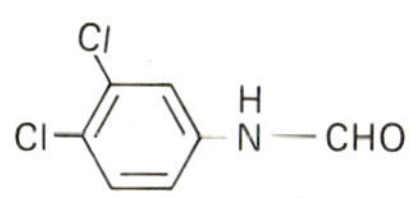

196

Kearney and Plimmer, 1972; Wallnofer et al., 1976), and 3,4-dichloroformanilide (196) were also found in the soil. Chlortoluron applied to soil results in the formation of monomethyl chlortoluron (Smith and Briggs, 1978).

4.1.13. Other Herbicides

The metabolism of a new herbicide oxadiazon (197) in soils under moist and flooded conditions was investigated by Ambrosi et al. (1977). It was observed that the metabolism of oxadiazon (197) proceeded by oxidation of the *tert*-butyl group to form a carboxylic acid derivative (199) and *O*-dealkylation of the isopropyl group to form a phenolic (198) and a methoxy (200) derivative (*Scheme 4.11*). A dealkylated derivative was also formed by oxidation. There was no evidence of either oxadiazon ring cleavage. Ring cleavage has been shown in rice plants (Hirata and Ishizuka, 1975).

4.2 INSECTICIDES

Soil insecticide metabolic research has received considerable attention in recent years. In general, organochlorine insecticides have received the most attention because they have been used longer and more extensively than organophosphorus and carbamate insecticides. Furthermore, the latter are usually degraded fairly rapidly in soils, in part by chemical reactions.

4.2.1. Organophosphates

Most of the organophosphates are readily degraded in soil mainly by hydrolytic and oxidative means. It has long been suspected that microorganisms are involved and actively participate in degrading organophosphorphates in soil.

4.2.1.1. Phosphorothioates

A comparison of autoclaved and nonautoclaved wet and dry soils indicated that the break down of the phosphorothioate parathion was brought about by the comparative numbers and metabolic activities of soil microorganisms (Lichtenstein and Schulz, 1964).

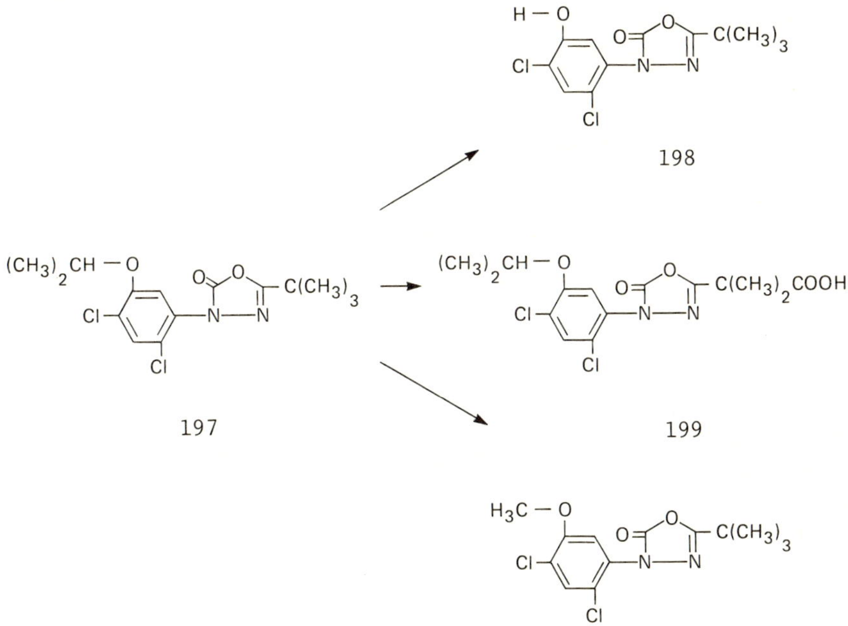

Scheme 4.11

Chemical sterilization of soil by sodium azide also decreased the degradation of parathion (Lichtenstein et al. 1968). Metabolism of parathion (69) in soil follows two pathways: hydrolysis to *p*-nitrophenol (201) and diethylthiophosphioric acid (202) and reduction to aminoparathion (203) (Lichtenstein and Schulz, 1964; Graetz et al. 1970; Sethunathan and Yoshida, 1973; Barik and Sethunathan, 1978a) (*Scheme 4.12*). In adopted mixed cultures, aminoparathion (203) produced from parathion (69) under low oxygen tension is hydrolyzed to *p*-aminophenol (204) and diethylthiophosphoric acid (202) (Munnecke and Hsieh, 1974). Paraoxon (205) formed in soil in small concentrations (Wolfe et al., 1973) can be detected in adapted mixed cultures (Munnecke and Hsieh,

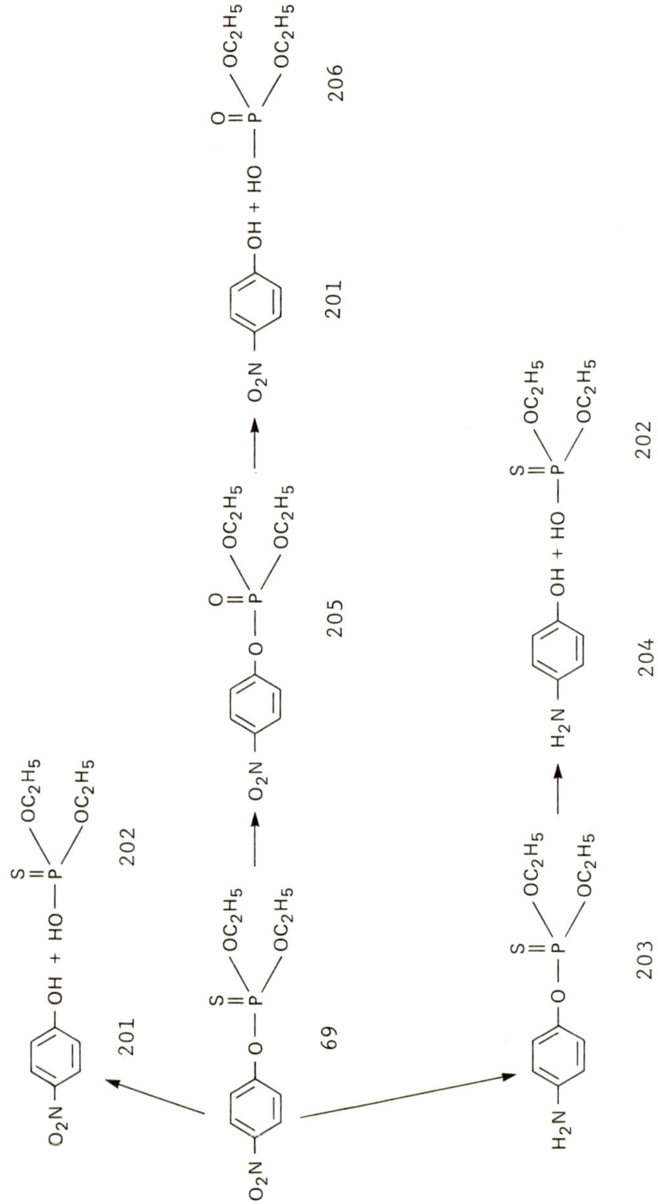

Scheme 4.12

1974). Complete disappearance of paraoxon (205) occurs by enzymatic hydrolysis to *p*-nitrophenol (201) and diethylphosphoric acid (206) (Munnecke and Hsieh, 1974).

Nitrophenol (201) released from parathion (69) is metabolized by bacteria isolated from flooded soils liberating nitrite and carbon dioxide (Barik et al., 1978a,b). However, the position and number of nitro substituents in the benzene ring will largely influence the degree of susceptibility of nitrophenols to biodegradation (Sethunathan et al., 1977). It was shown that a bacteria *Flavobacterium* sp. (Sethunathan and Yoshida, 1973), an alga, *Chlorella pyrenoidosa* (Mackiewicz et al., 1969), and a fungus, *Penicillium waksmani* isolated from an acid sulfate soil under flooded conditions degraded parathion (Rao and Sethunathan, 1974).

The major metabolite of fenitrothion by *B. subtilis* degradation is aminofenitrothion; other minor metabolites found are dimethylthiophosphoric acid and dimethyl fenitrothion (Miyamoto et al., 1966). The bacteria degrade aminofenitrothion slower than the parent compound, and desmethyl aminofenitrothion is identified as a metabolite. Methyl parathion is metabolized twice as fast as fenitrothion (Miyamoto et al., 1966). Recently, Spillner et al. (1979) showed that microbial degradation of fenitrothion in forest soils resulted in the formation of 3-methyl-4-nitrophenol, 3-methyl-4-nitroanisole and CO_2. These results were similar to those found in agricultural soils (Takimoto et al., 1976), but quite unlike the results obtained in flooded soil or mixed culture isolates from soil (Takimoto et al., 1976) where, in addition to 3-methyl-4-nitrophenol, such compounds as aminofenitrothion, 3-methyl-4-aminophenol and desmethylfenitrothion were observed. Spillner et al. (1979) suggested that the key intermediate in the degradation of 3-methyl-4-nitrophenol (and correspondingly fenitrothion) appears to be 2-methylhydroquinone which could undergo additional hydroxylation and/or oxidation of the methyl group, ortho-ring cleavage, and finally results in the formation of CO_2.

Microbiological degradation of diazinon in soil involves hydrolysis yielding *O,O*-diethyl phosphorothioate and 2-isopropyl-4-methyl-6-hydroxypyrimidine (Konrad et al., 1967). Evolution of $^{14}CO_2$ from ^{14}C-ring labeled and ^{14}C-ethyl labeled diazinon has been observed (Getzin and Rosefield, 1966; Getzin, 1967). Gunner

(1967) reported that species of *Pseudomonas*, *Arthrobacter* and *Streptomyces* degrade the hydrolytic products rather than the intact diazinon.

4.2.1.2. *Phosphorothiolothionates*

Malathion disappearance is much more rapid under nonsterile than under sterile conditions and the disappearance is stimulated by the various microbiological systems in the soil (Konrad et al., 1969; Walker and Stojanovic, 1973). Malathion is rapidly metabolized by a soil fungus, *Trichoderma viride*, and a bacterium *Pseudomonas* sp. isolated from soils which had received heavy application of the insecticide (Matsumura and Boush, 1966). Both the *Pseudomonas* bacterium and the *Trichoderma viride* fungus are most active in deethylating the carboxylic acid side chain of malathion (Matsumura and Boush, 1966). Walker and Stojanovic (1974) isolated an *Arthrobacter* sp. from soil that broke down malathion to malathion monoacid, malathion diacid, dimethyl phosphorodithioate, and dimethyl phosphorothioate.

Degradation of phorate (80) in the soil involves a rapid oxidation of the insecticide to phorate sulfoxide (207) and then a slow oxidation of the latter to phorate sulfone (208) (Menzer et al., 1970; Getzin and Shanks, 1970; Suett, 1971; Schulz et al., 1973; Lichtenstein et al., 1973) (*Scheme 4.13*). Ahmed and Casida (1958)

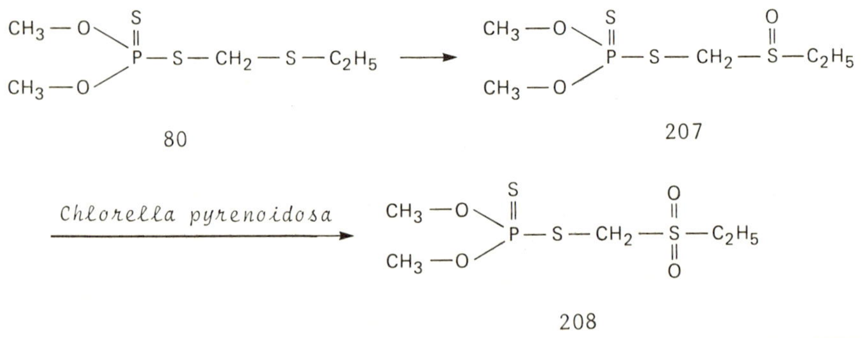

Scheme 4.13

studied the metabolism of phorate by various microorganisms. *Pseudomonas fluorescens* and *Thiobacillus thiooxidans* hydrolyzed phorate but no oxidation occurred. They also observed that with *Chlorella pyrenoidosa* phorate was oxidized to its sulfoxide (207), which was very stable to hydrolysis; the sulfoxide was converted slowly to its oxygen analogue (208).

Phorate sulfoxide (207) is the major metabolite present in water in submerged soils. However, in nonflooded soils, phorate sulfone (208) is the major metabolite (Walter-Echols and Lichtenstein, 1978). In subtropical soils, a rapid decrease in phorate concentration was accompanied by a concomitant increase in phorate sulfoxide and sulfone representing 18 and 74%, respectively, of total metabolites after 6 weeks (Talekar et al., 1977).

Microorganisms are partly responsible for dimethoate degradation in the soil (Getzin and Rosefield, 1968). The insecticide is metabolized in the soil to dimethoxon (Duff and Menzer, 1973). Species of *Aspergillus*, *Helminthosporium*, and *Streptomyces* utilize disulfoton as carbon and phosphorous sources. The major metabolites of disulfoton in the soil are disulfoton sulfoxide and disulfoton sulfone (Takase et al., 1972, 1973; Clapp et al., 1976).

4.2.1.3. Phosphates

Degradation of chlorfenvinphos, mevinphos and dichlorvos has been studied in soils (Beynon and Wright, 1967; Beynon et al., 1968; Getzin and Rosefield, 1968; Burns, 1971). Besides chlorfenvinphos (65), three major degradation products, desethylchlorfenvinphos (209), 2,4-dichloroacetophenone (213) and 1-(2,4-dichlorophenyl)-ethane-1-ol, (211) and traces of dichlorodiphenylethandiol (212) and dichlorophenacyl chloride (210) were found in the soil (Beynon and Wright, 1967) (*Scheme 4.14*). Dichlorvos and mevinphos are degraded more rapidly in nonsterile soil than in sterile soil (Getzin and Rosefield, 1968; Burns, 1971). Degradation of dichlorvos by *Pseudomonas melophthora*, bacteria (*Escherrichia*, *Protaminobacter*, and *Pseudomonas*) and *B. subtilis* has been reported (Yasuno et al., 1965; Boush and Matsumura, 1967; Hirakoso, 1969).

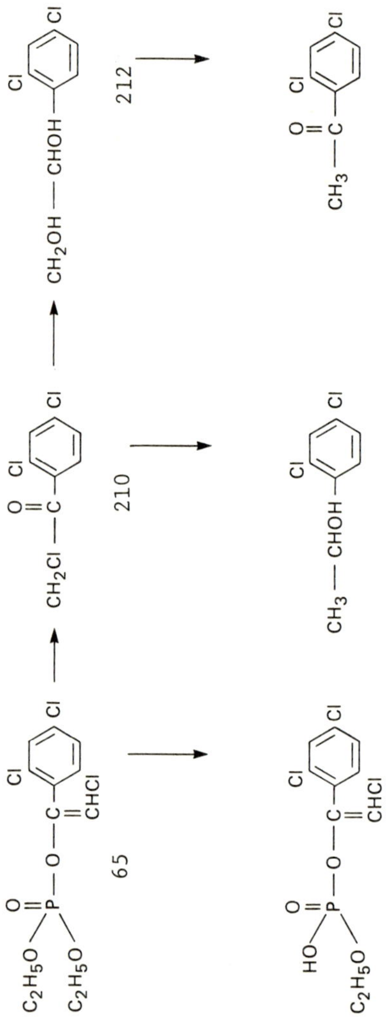

Scheme 4.14

4.2.1.4. Phosphonates

Microbial degradation is believed to be partially responsible for the disappearance of fonofos in the soil (Flashinski and Lichtenstein, 1974). The fungus *R. arrhizus* added to soil treated with ^{14}C-ethoxyl labeled fonfos degraded the insecticide during incubation. A significant amount of fonofoxon, as well as some water soluble labeled products were formed (Flashinski and Lichtenstein, 1974).

4.2.2. Carbamates

Carbamate pesticides have relatively short residual life times in the soil and are readily degraded by nontarget organisms. Caro et al. (1974) observed in a field study that during the first 40 days after an application of carbaryl to a silt loam soil no significant degradation of the insecticide took place due to a lag phase. However, about 135 days after the application 95% of carbaryl had disappeared. The soil fungus, *Aspergillus terreus* degrades carbaryl (86) to 1-naphthyl *N*-hydroxymethylcarbamate (214), 1-naphthyl carbamate (215), 4- (216), and 5-hydroxy-1-naphthyl methylcarbamates (217) (Liu and Bollag, 1971). 1-Naphthylcarbamate (215) is considered to be intermediary metabolite between 1-naphthyl *N*-hydroxymethylcarbamate (214) and 1-naphthol (218) (*Scheme 4.15*).

The main pathway of carbofuran degradation in soils is hydrolysis at the carbamate linkage. The carbamate moiety is degraded to CO_2, and the carbofuran phenol is rapidly bound to the soil. A gradual degradation of the carbofuran phenol follows with the release of CO_2 (Getzin, 1973). Liberation of $^{14}CO_2$ from ring labeled ^{14}C-carbofuran is taken as evidence of microbial degradation since microorganisms are implicated in ring cleavage of organic molecules. Among the microorganisms isolated from carbofuran amended soils, actinomycetes was particularly active in converting carbofuran to CO_2. In nonflooded soils, more rapid mineralization of ^{14}C-carbonyl labeled carbofuran to $^{14}CO_2$ occurs in nonsterile conditions (Williams et al., 1976). Venkateswarlu et al. (1977) demonstrated the involvement of microorganisms in the degradation of carbofuran in flooded soils. The insecticide is more rapidly hydrolyzed in rice soils under anaerobic conditions than under aerobic conditions. However, its hydrolysis products,

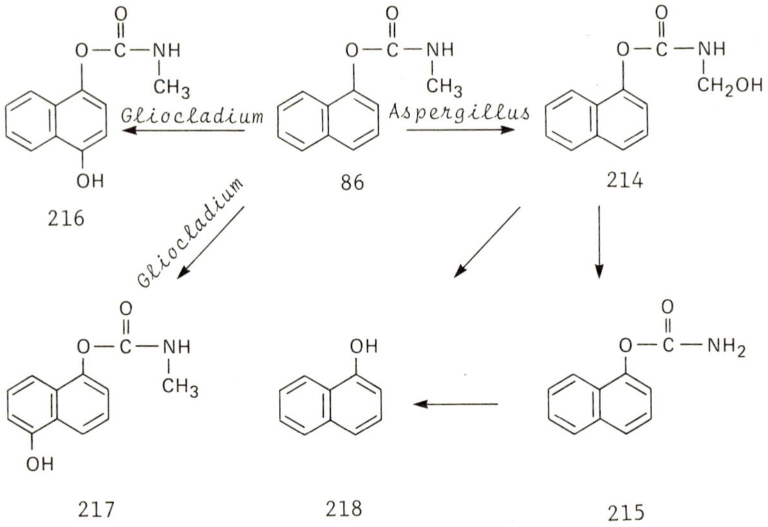

Scheme 4.15

carbofuran phenol and 3-hydroxycarbofuran, which resist further degradation under continued anaerobiosis, are rapidly degraded when the anaerobic system is returned to aerobic conditions (Venkateswarlu and Sethunathan, 1978). Siddaramappa et al. (1978) suggested that although hydrolysis of carbofuran in flooded soil was primarily chemical, degradation of carbofuran phenol was biological.

In soil after 12 weeks, aldicarb degrades rapidly with the formation of aldicarb sulfoxide as the major product (Coppedge et al., 1967). The sulfone, nitrile sulfoxide, oxime sulfoxide, and the oxime are also formed. Aldicarb sulfoxide and sulfone were the major solvent extractable metabolites in a laboratory study on the degradation of the insecticide in soils (Richey et al., 1977). Five soil fungi *Gliocladium catenulatum*, *Penicillium multicolor*, *Cunninghamella elegans*, *Rhizoctonia* sp. and *Trichoderma harzianum* metabolized aldicarb sulfoxide and the oxime and nitrile sulfoxide and traces of sulfones (Jones, 1976).

The carbamate insecticide, oxamyl undergoes biodegradation in the soil with less than 5% of the parent compound remaining one month after the application (Harvey and Han, 1978). The corresponding amino compound, methyl *N*-hydroxy-*N*,*N*-dimethyl-1-thiooxamimidate, is formed in the soils at the early stages, but this also decomposes rapidly. Microbial transformation is considered to be of major importance in determining the behavior of methomyl in soils (Fung and Uren, 1977).

4.2.3. Chlorinated Hydrocarbons

4.2.3.1. DDT and analogues

DDT (91) is stable in well aerated soils. However, when DDT amended soil is subjected to a reducing environment by either flooding or by maintaining oxygen free atmospheres (laboratory studies), the pesticide is dechlorinated to DDD (219) as the first intermediate (Guenzi and Beard, 1967). Adding organic materials to soils incubated anaerobically enhances the conversion rate of DDT (91) to DDD (219) (Guenzi and Beard, 1968; Parr et al., 1970; Parr and Smith, 1974).

The exact route by which DDT is fully degraded in soils is still not well understood. DDT (91) degrades much more readily in soils under anaerobic conditions to form DDD (219) but very slowly under aerobic conditions to yield DDE (220). The latter is formed by dehydrogenation of DDT (91) and is mediated by an enzyme system (Lipke and Kearns, 1960). DDD (219) and DDE (220) are not considered to be the sequential metabolites in the same pathway, but arise independently from DDT (91) (Plimmer et al., 1968). Break down of DDT (91) in vitro under anaerobic conditions by the bacterium *Enterobacter aerogenes* yields reduced dechlorinated compounds, oxidized derivatives, and ultimately DBP (229) (Barker and Morrison, 1965). Similar products are produced by microorganisms isolated from the soil (Chacko et al., 1966; Matsumura and Boush, 1968; Ko and Lockwood, 1968).

In soil under anaerobic conditions, minor metabolites such as DDA (226), dicofol (221), DBP (229), BA (230) and DDM (227), may also form (Guenzi and Beard, 1967). The majority of variants of the soil fungus *Trichoderma viride* produce DDD (219) and a dicofol like (221) compounds, whereas some variants exclusively produce DDA (221) or DDE (220) (Matsumura and Boush, 1968). This indicates the presence of entirely different metabolic pathways or variations in relative activities of enzyme systems in metabolizing DDT (91) among variants of the same species. The fungus *Fusarium oxysporum* can produce DDD (219) from either DDT (91) or DDE (220); the metabolic pathway then passes through DDMU (222), DDOH (225), and DDA (226) to DBH (228) (Engst and Kujawa, 1967). Incubation of DDT with *Aerobacter aerogenes* produces the metabolites DDMU (222), DDMS (223), DDNU (224) and DDOH (225) (Wedemeyer, 1968).

Very limited information is available on studies dealing with the metabolic fates of other DDT analogues by soil microorganisms. Menzie (1969) provided a list of microorganisms that can metabolize DDT (91) to DDD (219) and in some cases DDE (220). General pathways of DDT metabolism by soil microorganisms is shown in *Scheme 4.16*.

The rate of DDT degradation, and rates of formation and degradation of products produced are temperature dependent in flooded soil (Guenzi and Beard, 1976). The rate of DDT decomposition is at a maximum at 60°C with no degradation at 2°C. The anaerobic degradative pathway may be considered as DDT (91) → DDD (219) → DDMU (222). The accumulation of DDE (220) may result from the direct conversion of DDT (91), and after no more DDT is detected, the attained DDE (220) concentrations remain constant for each temperature (Guenzi and Beard, 1976).

4.2.3.2. Benzene hexachloride (γ-BHC)

Loss of BHC in the soil is attributed to a slow bacterial decomposition (Bradbury, 1963). γ-PCCH (2,3,4,5,6-pentachlorocyclohexene) is found in soils treated with lindane and the dehydrochlorination could be effected by *Bacillus cereus* isolated from the soil (Yule et al., 1967). Microbial degradation plays a significant role in the disappearance of lindane present in submerged soils (Raghu and MacRae, 1966). In a flooded clay loam soil, γ-HCH was degraded to γ-BTC and the conversion could be inhibited by the antibiotic sodium azide (Tsukano and Kobayashi, 1972). γ-HCH was dechlorinated to pentachlorocyclohexene by a *Clostridium* sp. isolated from a paddy soil. Lindane is converted to other isomers of hexachlorocyclohexane under submerged conditions (Newland et al., 1969). The α,β and δ isomers of benzene hexachloride are all rapidly degraded in flooded soil (MacRae et al., 1967).

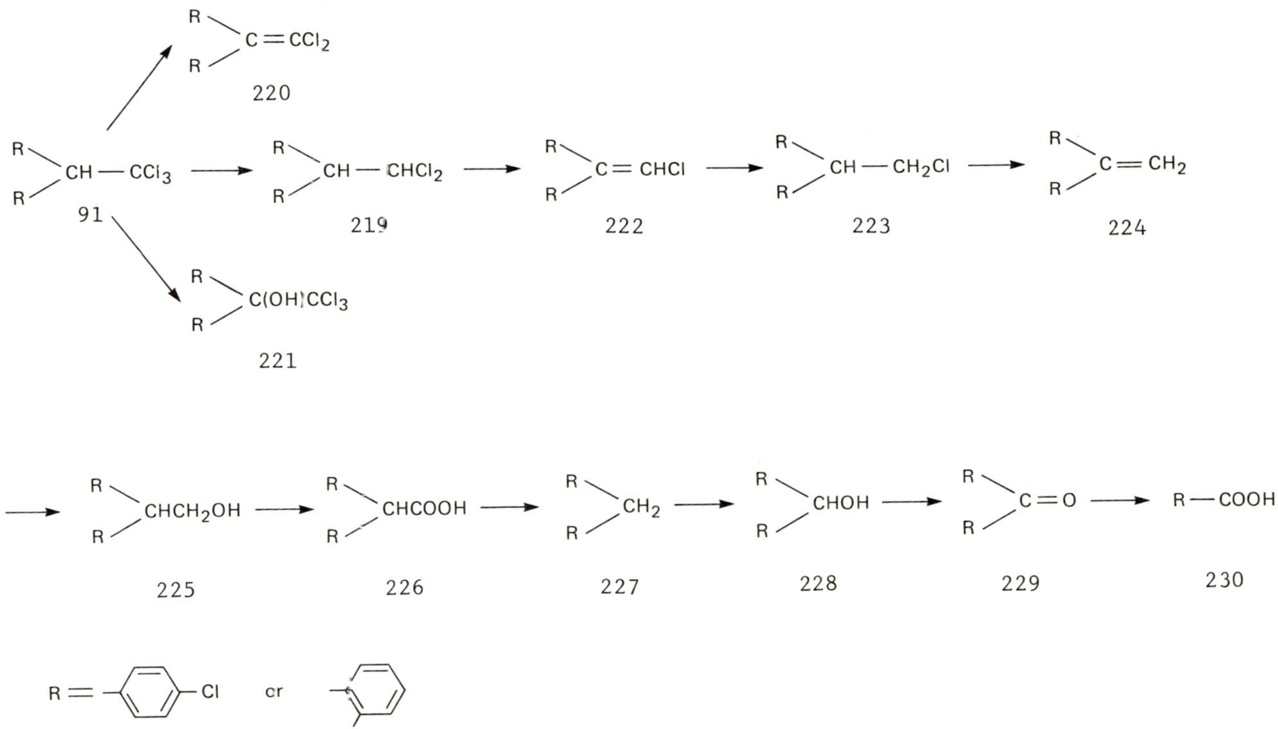
R
C=CCl2
R
220
R
CH—CCl3
R
91
R
CH—CHCl2
R
219
R
C=CHCl
R
222
R
CH—CH2Cl
R
223
R
C=CH2
R
224
R
C(OH)CCl3
R
221
R
CHCH2OH
R
225
R
CHCOOH
R
226
R
CH2
R
227
R
CHOH
R
228
R
C=O
R
229
R—COOH
230
R =
Cl
or
Cl

Scheme 4.16

The degradation of lindane is more rapid in submerged than in aerated moist soil (Kohnen et al., 1975). Yule et al. (1967) found γ-PCCH to be the only degradation product of lindane under aerobic conditions in a percolated and standing moist soil. Tsukano and Kobayashi (1972) detected only γ-3,4,5,6-tetrachlorocyclohexane and traces of γ-PCCH from lindane treated flooded soil. The microbial degradation of lindane in a sandy loam soil incubated for 6 weeks under flooded conditions resulted in the formation of metabolites γ-3,4,5,6-tetrachlorocyclohexane followed by γ-2,3,4,5,6-pentachlorocyclohex-1-ene and small amounts of 1,2,4-trichlorobenzene, 1,2,3,5-and/or 1,2,4,5-tetrachlorobenzene, and 1,2,3,4-tetrachlorobenzene (Mathur and Saha, 1975).

4.2.3.3. *Chlorinated cyclodienes*

Cyclodiene insecticides include such compounds as aldrin, dieldrin, heptachlor, isodrin, and endrin. The process of epoxidation of cyclodiene insecticides in sterile and nonsterile soils was suggested by Lichtenstein and Schulz (1960). The metabolic activity of soils involves the oxidation process leading to the formation of an epoxy ring from the unsaturated CH=CH bond of the unclorinated (or less chlorinated) ring (Gannon and Bigger, 1958; Keigemagi et al., 1958; Lichtenstein and Schulz, 1960; Bollen et al., 1958).

Laboratory cultures of *Aspergillus niger*, *A. flavus*, *Penicillium notatum*, and *P. chrysogenum* convert aldrin (95) to dieldrin (96) and several other metabolites (Korte et al., 1962). Tu et al. (1968) isolated *Trichoderma*, *Fusarium*, *Penicillium*, *Aspergillus*, *Nocardia*, *Streptomyces*, and *Micromonospora* species from a farm soil that had been previously shown to convert aldrin to dieldrin. Aldrin (95) is oxidized to dieldrin (96) in soils (Menzie, 1969) and epoxidation is mediated by soil microorganisms (Lichtenstein and Schulz, 1960).

Break down of dieldrin in the soil is very slow and the chlorinated ring moiety is very stable. Tu et al. (1968) observed that a number of *Trichoderma*, *Fusarium* and *Aspergillus* species isolated from aldrin treated soils were capable of degrading dieldrin. Matsumura and Boush (1967) isolated *T. viride*, *Pseudomonas* and *Bacillus* sp. from soil samples collected from places heavily contaminated with dieldrin and found that some of them degraded dieldrin (96) to a number of metabolites. (*E*)-Aldrin-diol (231) was the principal metabolite, but only 1 to 6% of the dieldrin was converted to the diol and six other water soluble compounds by these microorganisms. In a latter study, Matsumura and Boush (1968) observed that an isolate of *Trichoderma viride* from an Ohio orchard produced (*E*)-aldrin-diol and four other metabolites. Furthermore, it was observed that a *Pseudomonas* isolated from a soil sample taken from the cyclodiene manufacturing plant produced a different aldrin-diol plus one aldehyde and two ketoaldrins (232) (Matsumura et al., 1968) (*Scheme 4.17*).

More rapid degradation of endrin takes place in soils under flooded conditions than under nonflooded conditions (Gowda and Sethunathan, 1977). A culture of *A. flavus* metabolized endrin into two metabolites. The major metabolite was comparatively hydrophilic, and the minor metabolite was similar to ketoendrin (Korte, 1967).

Heptachlor (97) is oxidized to heptachlor epoxide (237) in soils (Gannon and Bigger, 1958; Young and Rawlins, 1958; Barthel et al., 1960; Lichtenstein and Schulz, 1960; Murphy and Barthel, 1960; Wilkinson et al., 1964). The conversion of 97 to 237 is caused by cultures of *Rhizopus*, *Fusarium*, *Pencillium*, *Trichoderma*, *Nocardia*, *Streptomyces*, *Bacillus*, and *Micromonospora* (Miles et al., 1969). Metabolism of heptachlor (97) involves chemical hydrolysis to 1-hydroxy-chlordene (233), which in subsequently microbially epoxidized to 1-hydroxy-2,3-epoxychlordene (234).

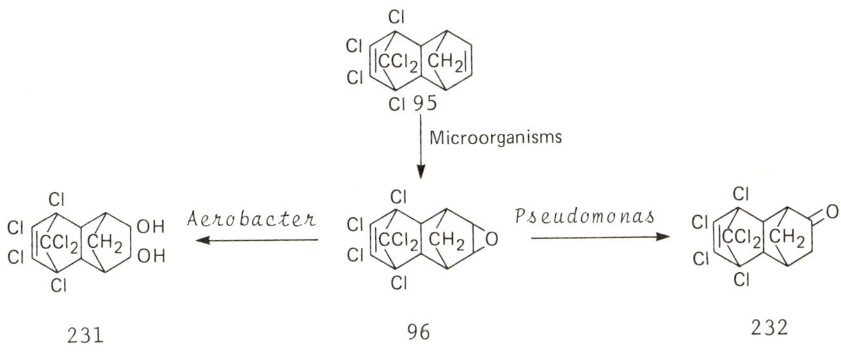

Scheme 4.17

Dechlorination of heptachlor (97) by microorganisms produces chlordene (235), which also undergoes microbial epoxidation to form the corresponding chlordene epoxide (236). A pathway showing heptachlor metabolism and chemical degradation in soils, according to Miles et al. (1969), is shown in *Scheme 4.18*.

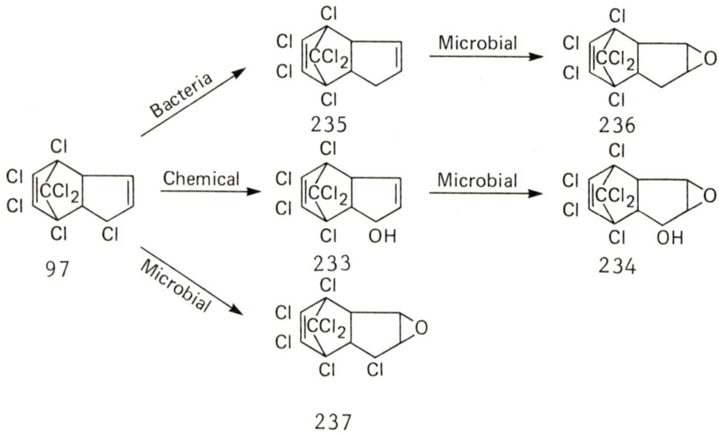

Scheme 4.18

4.2.4. Synthetic Pyrethroids

The degradation of the pyrethroid insecticide cypermethrin and the geometric isomers NRDC 160 (*Z*) and NRDC 159 (*E*) in three soils was reported by Roberts and Standen (1977a). The major degradative route in soils is hydrolysis of the ester linkage leading to the formation of 3-phenoxybenzoic acid and 3-(2,2-dichlorovinyl)-2,2-dimethylcyclopropanecarboxylic acid. Soil treated with the *Z*-isomer (NRDC 160) contains both *Z*- and *E*-isomer forms of the cyclopropanecarboxylic acid. A minor degradative route is ring hydroxylation of the insecticide to give an α-cyano-3-(4-hydroxyphenoxy) benzyl ester followed by hydrolysis of the ester bond. The pyrethroide inseciticde WL 41706, undergoes degradation by hydrolysis at the cyano group to form the amide and carboxylic acid analogues (Roberts and Standen, 1977b). However, the major degradative route is hydrolysis at the ester linkage leading initially to the formation of 3-phenoxybenzoic acid and 2,2,3,3,-tetramethylcyclopropanecarboxylic acid. Microbiological degradation of the *E* isomer of permethrin in soil occurs more rapidly than with the *Z* isomer (Kaufman et al., 1977). The major degradation mechanism of permethrin is hydrolysis to the dichlorovinyl acid and 3-phenoxybenzyl alcohol moieties. Further metabolism of both products results in the evaluation of CO_2 (Kaufman et al., 1977). Kaneko et al. (1978) also investigated the degradation of (+)-*E* and (+)-*Z* isomers of permethrin (98) in soil under laboratory conditions. The major degradation products in soil from both isomers were 3-(4-hydroxyphenoxy)benzyl-3-(2,2-dichlorovinyl)-2,2-dimethylcyclopropanecarboxylate (238), 3-phenoxybenzyl alcohol (239), 3-phenoxybenzoic acid (240), 3-(2,2-dichlorovinyl)-2,2-dimethylcyclopropanecarboxylic acid (241), and its hydroxylation derivative 242 (*Scheme 4.19*).

4.3. FUNGICIDES

The two mercury fungicides, Semesan® and Panogen® are degraded by soil microorganisms (Spanis et al., 1962). Semesan® is degraded by isolates of *Penicillium* sp. and *Aspergillus* sp. Panogen® is inactivated by several *Bacillus* sp. The degradation of PMA results in the formation of diphenylmercury as one of the major metabolites (Matsumura et al., 1971). Several other microorganisms convert phenylmercury to metallic mercury (Tonomura et al., 1968). Carbon mercury bond cleavage has been demonstrated to be the major reaction of a *Pseudomonad* on PMA (Furakawa et al., 1969). The metabolism of cacodylic acid proceeds via two routes in soils: C-As bond cleavage to arsenate and a carbon fragment under aerobic conditions, and reduction to alkylarsine under anaerobic condition (Kearney and Woolson, 1971).

PCNB (pentachloronitrobenzene) is reduced to pentachloroaniline by a large number of soil microorganisms (Menzie, 1969). The pentachloroaniline is stable in both moist and submerged soil (Ko and Farley, 1969).

In anaerobic soils, a loss of PCNB (243) occurs principally by conversion to pentachloroaniline (244) with some loss by volatilization and conversion to pentachlorothioanisole (245) and pentachlorophenol (246). Further degradation of pentachlorophenol (246)

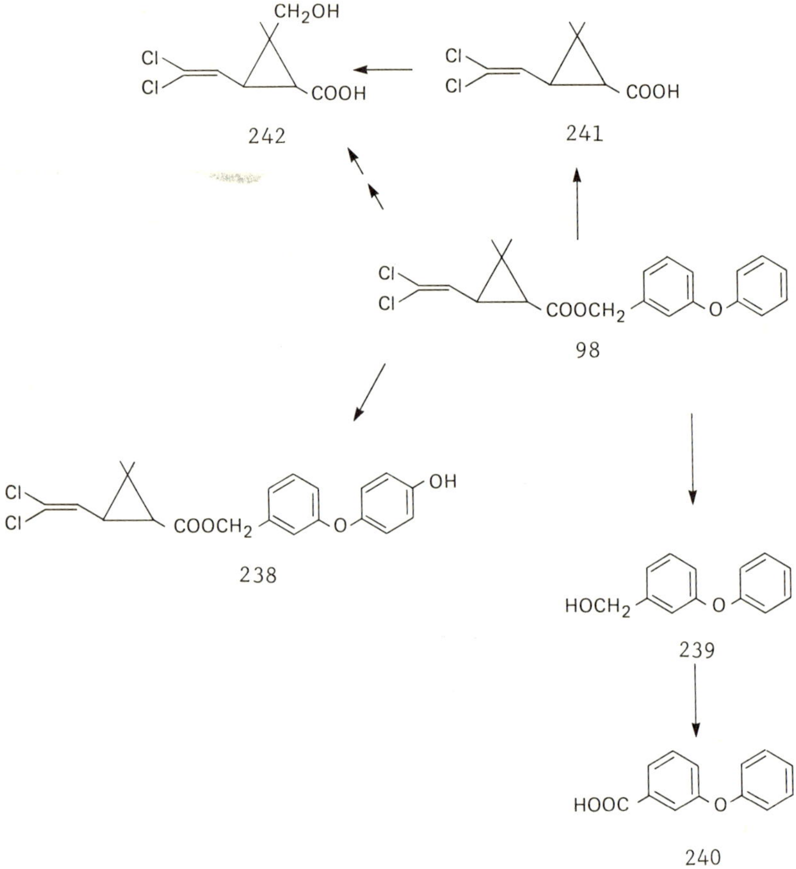

Scheme 4.19

results in the formation of 2,3,5,6-tetrachlorophenol (247), 2,3, 4,5-tetrachlorophenol (248) and 2,3,6-trichlorophenol (249) (Murthy and Kaufman, 1978; Murthy et al., 1979) (*Scheme 4.20*).

Metabolism of pentachlorophenol (246) by *Pseudomonas* sp. isolated from soil results in the release of CO_2 equivalent to approximately 50% of 246 added to bacterial suspension in one hour of incubation (Suzuki, 1977). The formation of metabolites tetrachlorocatechol

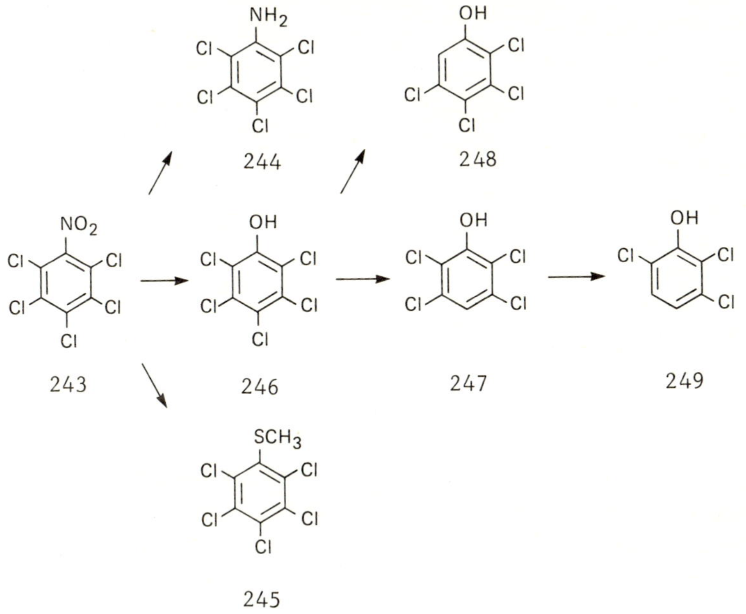

Scheme 4.20

(250) and tetrachlorohydroquinone (251) suggests that these products are intermediates prior to ring cleavage of pentachlorophenol (246).

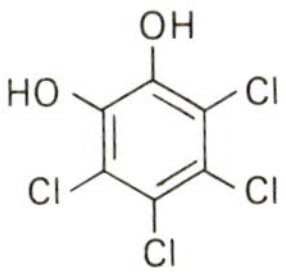

250

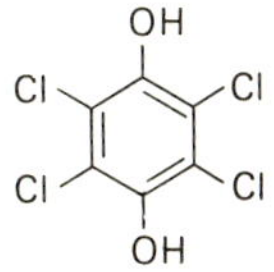

251

The soil fungi *Penicillium notatum*, *Glomerella cingulata* and *Fusarium roseum* produce CS_2 from thiram (104) (Sisler and Cox, 1951). The dimethyl dithiocarbamate ion (252), produced from thiram (104), may form amino acid adducts by action of soil micro-organisms. The half life of captan (105) in moist and dry silt loam soil was 3.5 and 50 days, respectively (Burchfield, 1959). Captan (105) produces tetrahydrophthalimide (253), thiophosgene (254), carbonyl sulfide (255) and H_2S by action of *Saccharomyces pastorianus* (Siegel and Sisler, 1968) (*Scheme 4.21*).

The systemic fungicide benomyl (110) is completely converted in soils to carbendazium (256) in a few hours (*Scheme 4.22*). Four species of bacteria and two of fungi have been isolated from soils that can effect the degradation of carbendazium (256) to non-fungicidal metabolites (Helweg, 1973). 2-Aminobenzimidazole (257) has also been isolated as a degradation product of benomyl (Kirkland et al., 1973; Baude et al., 1974). Helweg (1977) observed that carbendazium (256) added to the soil was slowly decomposed by microorganisms. 2-Aminobenzimidazole (257) was found as a degradation product although it appeared to be unstable in the soil,

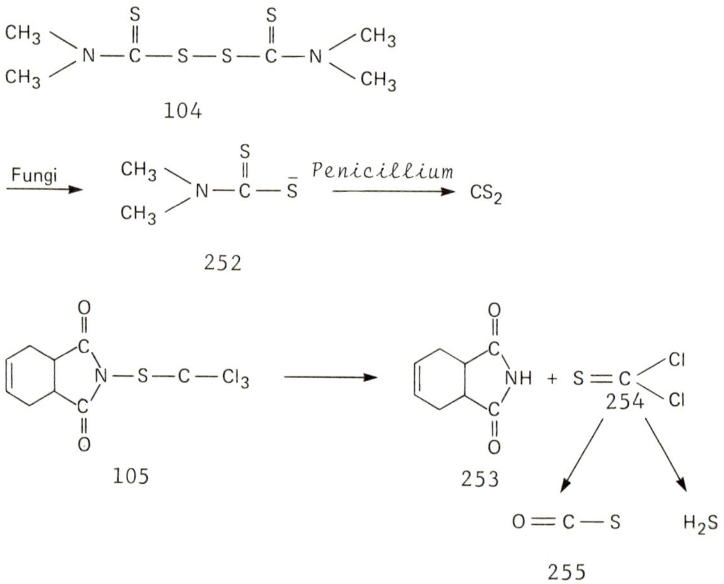

Scheme 4.21

decomposing rapidly after a lag period of about three weeks (*Scheme 4.22*). Baude et al. (1974) observed that in the field soil benomyl degraded to methyl 2-benzimidazole carbamate and 2-aminobenzimidazole.

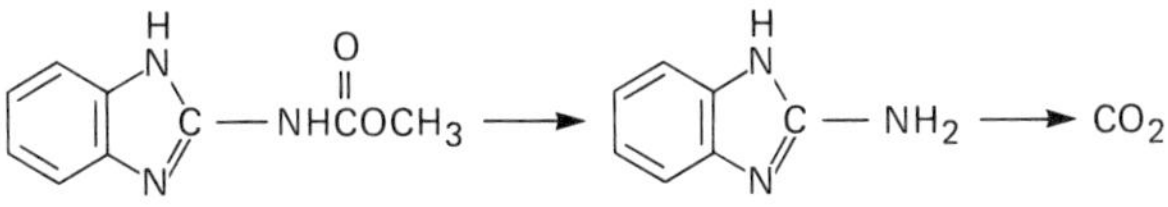

256 257

Scheme 4.22

4.4. FUMIGANTS

The soil fumigant dichloropropene mixture, (*Z*)- and (*E*)-1,3-dichloropropenes undergo hydrolysis in soil water slurries to the corresponding 3-chloroallyl alcohols (Castro and Belser, 1966). The metabolism of the isomeric 3-chloroallyl alcohols by a *Pseudomonas* sp. isolated from soil produces the corresponding 3-chloroacrylic acids. The latter are dehalogenated and converted to CO_2 (Belser and Castro, 1966). The degradation of 1,3-dichloropropenes and 3-chloroallyl alcohols in soils is mainly biological (Von Dijk, 1974). Roberts and Stoydin (1976) investigated the degradation in soil of 1,3 dichloropropene and 1,2-dichloropropane under laboratory and outdoor conditions. They observed that the conversion into the corresponding 3-chloroallyl alcohols and 3-chloroacrylic acids and the degradation products were also present in the soil in a bound form.

REFERENCES

Ahmed, M.K. and Casida, J.E., 1958. J. Econ. Entomol., 51: 59-63.

Alexander, M., 1965. Advanc. App. Microbiol., 7: 35-80.

Alexander, M. and Aleem, M.I.H., 1961. J. Agric. Food Chem., 9: 44-47.

Ambrosi, D., Kearney, P.C. and Macchia, J.A. 1977. J. Agric. Food Chem., 25: 868-872.

Baldwin, B.C., Bray, M.F. and Geogdegan, M.J., 1966. Biochem. J., 101: 15 p.
Barik, S. and Sethunathan, N., 1978a. J. Environ. Qual., 7: 346-348.
Barik, S. and Sethunathan, N., 1978b. J. Environ. Qual., 7: 349-352.
Barik, S., Siddaramappa, R. and Sethunathan, N., 1976. J. Microbiol. Serol., 42: 461-470.
Barker, P.S. and Morrison, F.O., 1965. Can. J. Zoo., 43: 652-654.
Barthel, W.F., Murphy, R.T., Mitchell, W.G. and Corely, C., 1960. J. Agric. Food Chem., 8: 445-447.
Baude, F.J., Pease, H.L. and Holt, R.F., 1974. J. Agric. Food Chem., 22: 413-418.
Beestman, G.B. and Deming, J.M., 1974. Agron. J. 66: 308-311.
Belser, N.O. and Castro, C.E., 1971. J. Agric. Food Chem., 19: 23-26.
Beynon, K.I. and Wright, A.N., 1967. J. Sci. Food Agric., 18: 143-150
Beynon, K.I. and Wright, A.N., 1968. J. Sci. Food Agric., 19: 718-722
Beynon, K.I., Edwards, M.J., Elgar, K. and Wright, A.N., 1968. J. Sci. Food Agric., 19: 302-307.
Bollag, J.M., Helling, C.S. and Alexander, M., 1967. Appl. Microbiol., 15: 1393-1398.
Bollag, J.M., Helling, C.S. and Alexander, M., 1958. J. Agric. Food Chem., 16: 826-829.
Bollen, W.P., Roberts, J.E. and Morrison, H.E., 1968. J. Econ. Entomol., 51: 214-219.
Bordeleau, L.M. and Bartha, R., 1971. Soil Biol. Biochem., 3: 281-284.
Boush, G.M. and Matsumura, F., 1967. J. Econ. Entomol., 60: 918-920.
Bradbury, F.R., 1963. Ann. Appl. Biol., 52: 361-370.
Burchfield, H.P., 1959. Contrib. Boyce Thompson Inst., 20: 205-215.
Burns, R.G., 1971. Bull. Environ. Contam. Toxicol., 6: 316-321.
Byrde, R.J.W. and Woodcock, D., 1957. Biochem. J., 65: 682-684.
Caro, J.H., Freeman, H.P. and Turner, B.C., 1974. J. Agric. Food Chem., 22: 860-863.
Chacko, C.I., Lockwood, J.L. and Zabik, M., 1966. Science, 154: 893-895.
Castro, E.C. and Belser, N.O., 1966. J. Agric. Food Chem., 14: 69-70.
Chisaka, H. and Kearney, P.C., 1970. J. Agric. Food Chem., 18: 854-858.
Clapp, D.W., Naylor, D.V. and Lewis, G.C., 1976. J. Environ. Qual. 5: 207-210.
Coppedge, J.R., Lindquist, D.A., Bull, D.L. and Dorough, H.W., 1967. J. Agric. Food Chem., 15: 902-910.
Corbins, F.T. and Upchurch, R.P., 1967. Weeds, 15: 370-377.
Couch, R.W., Granhich, J.V. Davis, D.E. and Funderburk, H.H., Jr. 1965. Proc. S. Weed Conf., 18: 623.
Dalton, R.L., Evans, A.W. and Rhodes, R.C., 1966. Weeds, 14: 31-33.
Dewey, O.R., Lindsay, R.V. and Hartley, G.S., 1962. Nature, 195: 1232.
Duff, W.G. and Menzer, R.E. 1973. Environ. Entomol., 2: 309-318.
Duxbury, J.M., Tiedje, J.M., Alexander, M. and Dawson, J.E., 1970. J. Agric. Food Chem., 18: 199-201.
Engelhardt, G., Wallnofer, P.R. and Plapp, R., 1972. Appl. Microbiol., 23: 664-666.
Engst, R. and Kujawa, M. 1967. Nahrung, 11: 751-760.
Esser, H.O., Dupuis, G., Ebert, E., Marco, G.J. and Vogel, C.,

1975. In: P.C. Kearney and D.D. Kaufman (Editors), Herbicides, Vol. 1. Dekker, New York, N.Y. pp. 129-208.
Fang, S.C., 1975. In: P.C. Kearney and D.D. Kaufman (Editors), Herbicides, Vol. 1. Dekker, New York, N.Y., pp. 323-349.
Faulkner, J.K. and Woodcock, D., 1965. Nature, 203: 865-867.
Fernley, H.N. and Evans, W.C., 1959. Biochem. J., 73: 22 p.
Flashinski, S.J., Lichtenstein, E.P., 1974. Can. J. Microbiol., 20: 871-875.
Funderburk, H.H. Jr., and Bozarth, G.A., 1967. J. Agric. Food Chem., 15: 563-567.
Fung, K.K.H. and Uren, N.C., 1977. J. Agric. Food Chem., 25: 966-969.
Furukawa, F., Suzuki, T. and Tonomura, K., 1969. Agric. Biol. Chem., 33: 128-130.
Gannon, N. and Bigger, J.H., 1958. J. Econ. Entomol., 51: 1-2.
Gaunt, J.K. and Evans, W.C., 1961. Biochem. J., 79: 25P-26P.
Gaunt, J.K. and Evans, W.C., 1971. Biochem. J., 122: 519-526; 533-542.
Geissbuhler, H., Martin, H. and Voss, G., 1975. In: P.C. Kearney and D.D. Kaufman (Editors), Herbicides, Vol. I, Dekker, New York, N.Y., pp. 209-291.
Geissbuhler, H., Haselback, C., Aebi, H. and Ebner, L., 1963. Weed Res., 3: 277-297.
Getzin, L.W., 1967. J. Econ. Entomol., 60: 505-558.
Getzin, L.W., 1973. Environ. Entomol., 2: 461-467.
Getzin, L.W. and Rosefield, I., 1966. J. Econ. Entomol., 50: 512-516.
Getzin, L.W. and Rosefield, I., 1968. J. Agric. Food Chem., 16: 598-601.
Getzin, L.W. and Shanks, C.H. Jr., 1970. J. Econ. Entomol., 63: 52-58.
Golab, T., Althaus, W.A. and Wooten, H.L. 1979. J. Agric. Food Chem., 27: 163-179.
Gowda, T.K.S. and Sethunathan, N., 1977. Soil Sci. 124: 5-9.
Graetz, D.A., Chesters, G., Daniel, T.C., Newland, L.W. and Lee, G.B., 1970. J. Water Pollnt. Control Fed., 42: 76-94.
Grover, R., 1967. Weed Res., 7: 61-67.
Guenzi, W.D. and Beard, W.E., 1967. Science, 156: 1116-1117.
Guenzi, W.D. and Beard, W.E., 1968. Soil Sci. Soc. Am. Proc., 32: 522-524.
Guenzi, W.D. and Beard, W.E., 1976. J. Environ. Qial., 5: 391-394.
Gunner, H.B., 1967. In: Coop. Reg. Proj. NE-53 CSRS, U.S. Dept. Agriculture, Washington, D.C. pp. 1-14.
Gutenmann, W.H. and Lisk, D.J., 1964. J. Agric. Food Chem., 12: 322-323.
Gutenmann, W.H., Loos, M.A., Alexander, M. and Lisk, D.J., 1964. Soil Sci. Soc. Am. Proc., 28: 205-207.
Hargrove, R.S. and Merkel, M.G., 1971. Weed Sci., 19: 652-654.
Harris, C.I., Kaufman, D.D., Sheets, T.J., Nash, R.G. and Kearney, P.C., 1968. Residue Rev., 8: 1-55.
Harvey, J. Jr. and Han, J.C.Y., 1978. J. Agric. Food Chem., 26: 536-541.
Helling, C.S., 1976. J. Environ. Qual., 5: 1-14.
Helling, C.S., Bollag, J.M. and Dawson, J.E., 1968. J. Agric. Food Chem., 16: 538-539.
Helling, C.S., Kearney, P.C. and Alexander, M., 1971. Advan. Agron., 23: 147-240.
Helweg, A., 1973., Tijdsskr. Planteavl., 77: 232-243.
Helweg, A., 1977. Pestic. Sci., 8: 71-78.

Hirakoso, S. 1969. Jpn. J. Exp. Med., 39: 17-20.
Hirata, H. and Ishizuka, K., 1975. Agric. Biol. Chem., 39: 1447-1454.
Hock, W.K. and Sisler, H.D., 1969. J. Agric. Food Chem., 17: 123-128.
Horvath, R.S., 1971. J. Agroc. Food Chem., 19: 291-293.
Jones, A.S., 1976. J. Agric. Food Chem., 24: 115-117.
Kaneko, H., Ohkawa, H. and Miyamoto, J., 1978. J. Pestic. Sci., 3: 43-51.
Kaufman, D.D., 1967. J. Agric. Food Chem., 15: 582-591.
Kaufman, D.D., 1970. In: Pesticide in the Soil: Ecology, degradation, movement, Int. Symp., Michigan State Univ., East Lansing, pp. 73-86.
Kaufman, D.D., 1974. In: W.D. Guenzi (Editor), Pesticide in Soil and Water, Soil Sci. Soc. Am. Inc. Publ., Madison, Wisc., pp. 133-202.
Kaufman, D.D. and Blake, J., 1970. Soil Biol. Biochem., 2: 73-80.
Kaufman, D.D. and Blake, J., 1973. Soil Biol. Biochem., 5: 297-308.
Kaufman, D.D. and Kearney, P.C., 1965. Appl. Microbiol., 13: 443-446.
Kaufman, D.D. and Kearney, P.C., 1970. Residue Rev. 32: 235-265.
Kaufman, D.D. and Plimmer, J.R., 1971. Weed Sci. Soc. Am. Abstr. Soc., 1971, No. 35, p. 18.
Kaufman, D.D. and Plimmer, J.R., 1972. In: R. Mitchell (Editor), Water Pollution Microbiology, Willy Interscience, New York, N.Y., pp. 173-203.
Kaufman, D.D., Kearney, P.C. and Sheets, T.J., 1963. Science, 142: 405-406.
Kaufman, D.D., Kearney, P.C. and Sheets, T.J., 1965. J. Agric. Food Chem., 13: 238-242.
Kaufman, D.D., Haynes, S.C., Jordan, E.G. and Kayser, A.J., 1977. ACS Symp. Ser., 42: 147-161.
Kearney, P.C., 1965. J. Agric. Food Chem., 13: 561-564.
Kearney, P.C. and Plimmer, J.R., 1972. J. Agric. Food Chem., 20: 584-585.
Kearney, P.C. and Woolson, E.A., 1971. Abstr. 162nd Meet. Am. Chem. Soc. Div. Pestic. Chem., No. 29.
Kearney, P.C., Kaufman, D.D. and Sheets, T.J., 1965. J. Agric. Food Chem., 13: 369-372.
Kearney, P.C., Plimmer, J.R. and Guardia, F.B., 1969. J. Agric. Food Chem., 17: 1418-1419.
Kearney, P.C., Smith, R.J., Plimmer, J.R. and Guardia, F.S., 1970. Weed Sci., 18: 464-466.
Kesner, C.D. and Ries, S.K., 1967. Science, 155: 210-211.
Khan, S.U. and Marriage, P.B., 1977. J. Agric. Food Chem., 25: 1408-1413.
Keigemagi, V., Morrison, H.E., Roberts, J.E. and Bollen, W.B., 1958, J. Econ. Entomol., 51: 198-204.
Kirkland, J.J., Holt, R.F. and Pease, H.L., 1973. J. Agric. Food Chem., 21: 368-371.
Ko, W.H. and Farley, J.D., 1969. Phytopathology, 59: 64-67.
Ko, W.H. and Lockwood, J.L., 1968. Can. J. Microbiol. 14: 1069-1073.
Kohnen, R., Haider, K. and Jagnow, G., 1975. Environ. Qual. Safety Suppl. III, 222-255.
Konrad, J.G., Armstrong, D.E. and Chesters, G., 1967. Agron. J., 59: 591-594.

Kondrad, J.G., Chesters, G. and Armstrong, D.E., 1969. Soil Sci. Soc. Am. Proc., 33: 259-262.
Korte, F., 1967. Proceedings of the Commission on Terminal Residues and of the Commission on Residue Analysis, IUPAC, Vienna.
Korte, F., Ludwig, G. and Vogel, J., 1962. Justus Liebig's Ann. der Chemie, 656: 135-140.
Laanio, T.L., Kearney, P.C. and Kaufman, D.D., 1973. Pestic. Biochem. Physiol, 3: 271-277.
Laveglia, J. and Dahm, P.A., 1977. Ann. Rev. Entomol., 22: 483-513.
Lichtenstein, E.P. and Schulz, K.R., 1960. J. Econ. Entomol., 53: 192-197.
Lichtenstein, E.P. and Schulz, K.R., 1964. J. Econ. Entomol., 57: 618-627.
Lichtenstein, E.P., Fuhremann, T.W. and Schulz, K.R., 1968. J. Agric. Food Chem., 16: 870-873.
Lichtenstein, E.P., Fuhremann, T.W., Schulz, K.R. and Liang, T.T., 1973. J. Econ. Entomol., 66: 863-866.
Lipke, H. and Kearns, C.W., 1960. Advan. Pest. Contr. Res., 3: 253-287.
Liu, S.Y., and Bollag, J.M., 1971. Pestic. Biochem. Physiol., 1: 366-372.
Loos, M.A., 1975. In: P.C. Kearney and D.D. Kaufman (Editors), Herbicides, Vol. 1., Dekker, New York, N.Y., pp. 1-128.
Mackiewicz, M., Duebert, K.H., Gunner, H.B. and Zuckerman, B.M., 1969. J. Agric. Food Chem., 17: 129-130.
MacRae, I.C. and Alexander, M., 1964. Agron. J., 56: 91-92.
MacRae, I.C. and Alexander, M., 1965. J. Agric. Food Chem., 13: 72-76.
MacRae, I.C., Raghu, K. and Castro, T.F. 1967. J. Agric. Food Chem., 15: 911-914.
Mathur, S.P. and Saha, J.G., 1975. Soil Sci., 120: 301-307.
Matsumura, F. and Boush, G.M., 1966. Science, 153: 1278-1280.
Matsumura, F. and Boush, G.M., 1967. Science, 156: 959-961.
Matsumura, F. and Boush, G.M., 1968. J. Econ. Entomol., 61: 610-612.
Matsumura, F. and Boush, G.M., 1971. In: A.D. McLaren and J. Skujuns (Editors), Soil Biochemistry, Dekker, New York, N.Y., pp. 320-336.
Matsumura, F., Boush, G.M. and Tai, A., 1968. Nature, 29: 965-967.
Matsumura, F., Gotoh, Y. and Boush, G.M., 1971. Science, 173: 49-51.
Meikle, R.W., Williams, E.A. and Redemann, C.T., 1966. J. Agric. Food Chem., 14: 384-387.
Menzer, R.E., Fontanilla, E.L. and Ditman, L.P., 1970. Bull. Environ. Contam. Toxicol., 5: 1-5.
Menzie, C.M., 1969. U.S. Fish Wildlife Serv.; Special Report, 127.
Messersmith, C.G., Burnside, O.C. and Lavy, T.L., 1971. Weed Sci., 19: 285-290.
Miles, J.R.W., Tu, C.M. and Harris, C.R., 1969. J. Econ. Entomol., 62: 1334-1338.
Miyamoto, J., Kitagawa, K. and Sato, Y., 1966. Jpn. J. Exp. Med., 36: 211-225.
Moshier, L.J. and Penner, D., 1978. Weed Sci., 26: 686-691.
Muir, D.C.G. and Baker, B.E., 1978. Weed Res., 18: 111-120.
Munnecke, D.M. and Hsieh, D.P.H., 1974. App.. Environ. Microbiol., 31: 63-69.
Murphy, R.T. and Barthel, W.F., 1960. J. Agric. Food Chem., 8: 422-445.
Murray, D.S. and Reick, W.L., 1968. Agron. Abstr., 60: 95.
Murthy, N.B.K. and Kaufman, D.D., 1978. J. Agric. Food Chem., 26: 1151-1156.

Murthy, N.B.K., Kaufman, D.D. and Fries, G.F., 1979. J. Environ. Sci. Health, B14: 1-14.
Newland, L.W., Chesters, G. and Lee, G.B., 1969. J. Water Pollut. Contr. Fed., 41: R174-R188.
Owens, R.G., 1960. Develop. Ind. Microbiol., 1: 187-205.
Parr, J.F. and Smith, S., 1974. Soil Sci., 118: 45-52.
Parr, J.F., Willis, G.H. and Smith, S., 1970. Soil Sci. 110: 306-312.
Plimmer, J.R., kearney, P.C. and Von Endt, D.W., 1968. J. Agric. Food Chem., 16: 594-597.
Plimmer, J.R., Kearney, P.C. and Chisaka, H., 1970a. Weed Sci. Soc. Am. Abst. No. 167.
Plimmer, J.R., Kearney, P.C., Chisaka, H., Young, J.B. and Klingebiel, U.I., 1970b. J. Agric. Food Chem., 18: 859-861.
Probst, G.W., Golab, T. and Wright, W.L., 1975. In: P.C. Kearney and D.D. Kaufman (Editors), Herbicides, Vol. 1, Dekker, New York, N.Y., pp. 453-500.
Raghu, K. and MacRae, I.C., 1966. Science, 154: 263-264.
Rao, A.V. and Sethunathan, N., 1974. Arch. Microbiol., 97: 203-208.
Rauser, W.E. and Switzer, C.M., 1962. Weeds, 10: 62-64.
Reid, J.J., 1963. Science for the Farmers (Penn. State Univ., Agr. Exp. Sta.), Vol. X.
Richey, F.A. Jr., Bartley, W.J. and Sheets, K.P., 1977. J. Agric. Food Chem., 25: 47-51.
Roberts, T.R. and Stoydin, G., 1976. Pestici. Sci. 7: 325-335.
Roberts, T.R. and Standen, M.E., 1977a. Pestic. Sci., 8: 305-319.
Roberts, T.R. and Standen, M.E., 1977b. Pestic. Sci., 8: 600-610.
Rueppel, M.L., Brightwell, B.B., Schaefer, J. and Marvel, J.T., 1977. J. Agric. Food Chem., 25: 517-528.
Schroeder, M., 1970. Weed Sci. Soc. Am. Abstr., No. 157.
Schulz, K.R., Lichtenstein, E.P., Fuhremann, T.W. and Liang, T.T., 1973. J. Econ. Entomol., 66: 873-875.
Sethunathan, N. and Yoshida, T., 1973. J. Agric. Food Chem., 21: 504-506.
Sethunathan, N., Siddaramappa, R., Rajaran, K.P., Barik, S. and Wahid, P.A., 1977. Residue Rev. 68: 91-122.
Sheets, T.J., 1959. Weeds, 7: 442.
Sheets, T.J., Smith, J.W. and Kaufman, D.D., 1968. Weed Sci., 16: 217.
Siddaramappa, R., Tirol, A.C., Seiber, J.N., Heinriches, E.A. and Watanab, I., 1978. J. Environ. Sci. Health, B13: 369-380.
Siegel, M.R. and Sisler, H.D., 1968. Phytopathology 58: 1123-1133.
Sisler, H.D. and Cox, C.E. 1951. Phytopathology, 41: 465.
Smith, R.A., 1972. Meet., Am. Chem. Soc. Abst. No. 27.
Smith, A.E. and Phillips, D.V., 1976. J. Agric. Food Chem., 24: 294-296.
Smith, A.E. and Briggs, G.G., 1978. Weed Res., 18: 1-7.
Smith, J.W. and Sheets, T.J., 1967. Weed Sci. Soc. Am. Abst., p. 76.
Spanis, W.C., Munnecke, D.E. and Solberg, R.A., 1962. Phytopathology, 52: 455-462.
Spillner, C. J. Jr., DeBaun, J.R. and Menn, J.J., 1979. J. Agric. Food Chem., 27: 1054-1060.
Sprankle, P., Sanberg, C.L., Meggitt, W.F. and Penner, D., 1978. Weed Sci. 26: 673-674.
Suett, D.L., 1971. Pestic. Sci. 2: 105-112.
Suzuki, T., 1977. J. Environ. Sci. Health, B12: 113-127.
Takase, I., Tsuda, H. and Yoshimoto, Y., 1972. Pflanzenschutz-Nachr, 25: 43-63.
Takase, I., Nakamura, H., Kobayashi, M., Tsuboi, A. and Wakabayshi, S., 1973. Pesticide Abstracts 7: 223-224.

Takimoto, Y., Hiroto, M., Inui, H. and Miyamoto, J., 1976. J. Pestic. Sci., 1: 131-139.

Talekar, N.S., Sun, L.T., Lee, E.M. and Chen, J.S., 1977. J. Agric. Food Chem., 25: 348-352.

Taylor, R.M.S., 1972. Thesis, Michigan State University, E. Lansing, Mich.

Tiedje, J.M. and Alexander, M., 1969. J. Agric. Food Chem., 17: 1080-1084.

Tiedje, J.M. and Hagedorn, M.L., 1975. J. Agric. Food Chem., 23: 77-81.

Tiedje, J.M., Duxbury, J.M., Alexander, M. and Dawson, J.E., 1969. J. Agric. Food Chem., 1021-1026.

Tonomura, K., Maeda, K., Futai, F. Nakagaini, T. and Yamada, M., 1968. Nature, 217: 644-646.

Torgeson, D.C. and Mee, H., 1967. Proc. Northwest. Weed Control Conf., 584-587.

Tsukano, Y. and Kobayashi, A., 1972. Agric. Biol. Chem., 36: 166-167.

Tu, C.M., Miles, J.R.W. and Harris, C.R., 1968. Life Sci., 7: 311-323.

Von Dijk, H. 1974. Agro-Ecosystems, 1974, 1: 193-204.

Von Endt, D.W., Kearney, P.C. and Kaufman, D.D., 1968. J. Agric. Food Chem., 16: 17-20.

Venkateswarlu, K., gowda, T.K.S. and Sethunathan, N., 1977. J. Agric. Food Chem., 25: 533-536.

Venkateswarlu, K. and Sethunathan, N., 1978. J. Agric. Food Chem., 26: 1148-1151.

Viswanathan, R., Scheunert, I., Kohli, J., Klein, W. and Korte, F. 1978. J. Environ. Sci. Health, B13: 243-259.

Walker, W.W. and Stojanovic, B.J., 1973. J. Environ. Quality, 2: 229-232.

Walker, W.W. and Stojanovic, B.J., 1974. J. Environ. Quality, 3: 4-13.

Wallnofer, P.R., 1969. Weed Res., 9: 333-339.

Wallnofer, P.R., Engelhardt, G. and Fuchsbichler, G., 1976. Bayer. Landw. Jb., 53: 309-317.

Walter-Echols, G. and Lichtenstein, E.P., 1978. J. Agric. Food Chem., 26: 599-604.

Wedemeyer, G., 1967. Appl. Microbiol., 15: 569-574.

Wheeler, W.B., Stratton, G.D., Twilley, R.R., Ou, Li-Tse, Carlson, D.A. and Davidson, J.M., 1979. J. Agric. Food Chem., 27: 702-706.

Wildung, R.E., Chesters, G. and Armstrong, D.E., 1968. Weed Res., 8: 213-225.

Wilkinson, A.T.S., Finlayson, D.G. and Morley, H.V., 1964. Science, 143: 681-682.

Williams, I.H., Pepin, H.S. and Brown, M., 1976. J. Bull. Environ. Contam. Toxicol., 15: 244-249.

Wolf, D.C. and Martin, J.P., 1975. J. Environ. Qual., 4: 134-139.

Wolf, H.R., Staiff, D.C., Armstrong, J.F. and Comer, S.W., (1973). Bull. Environ, Contam. Toxicology, 10: 1-9.

Woolson, E.A. and Kearney, P.C., 1973. Environ. Sci. Technol., 7: 47.

Wright, K.A. and Cain, R.B., 1972. Biochem. J., 128: 543-599; 561-569.

Wurzer, V.B. and Corbin, F.T., 1968. Z. Pflanzenkrankeiten und Pflanzenschutz, 75: 175-185.

Yasuno, M., Hirakoso, S., Sasa, M. and Uchida, M., 1965. Jpn. J. Exp. Med., 35: 545-563.

Young, W.R. and Rawlins, W.A., 1958. J. Econ. Entomol., 51: 11-18.

Youngson, C.R., Goring, C.A.I., Meikle, R.W., Scott, H.H. and Griffith, J.D., 1967. Down to Earth, 23: 2-11.

Yule, W.N., Chiba, M. and Morely, H.V., 1967. J. Agric. Food Chem., 15: 1000-1004.

Chapter 5

OCCURRENCE AND PERSISTENCE OF PESTICIDE RESIDUES IN SOIL

A pesticide residue in soil may be considered as any substance or mixture of substances in or on soil resulting from the use of a pesticide. This includes any derivatives, such as conversion and degradation products, reaction products, metabolites and impurities. Only a portion of the pesticide residues found in soil result from direct application. Other important sources of pesticide residues in soil are from spray fall out, in rain and dust, and from crop and animal remains. Sprays applied to crop foliage do not always reach their target. It has been estimated that as much as 50% of the pesticide applied to crop foliage reaches the soil, either as spray drift or run off from the leaves or on leaves that fall to the ground. In orchards, all the pesticides are applied to the foliage, thus the soil residues are due to foliar spraying and not from their direct application to the soil. A large proportion of the residues in soils may also originate from aerial spraying of crops and forests.

Atmosphere may contain residues of pesticides that are likely to be added to soil with rainfall. Residues have been reported in rain, air and dust. It is likely that these residues originate from spray drift or by volatilisation from soil to water. It is assumed that the residues become concentrated on to particulate matter or in moisture drops and fall on soil either with dust or rain. However, the amounts that reach in this way are unlikely to be large. Wheatley and Hardman (1965) concluded that no significant increase in the contamination of agricultural land seems likely to arise from the amounts of organochlorine insecticide residues they found in rain water.

Pesticides may also reach the soil from plant or animal remains which become incorporated with the soil. Sufficient data are available in the literature showing that small quantities of pesticide residues are taken up from soils in the tissue of plants.

These residues ultimately reach the soil when crops are ploughed in the field. Some pesticides are concentrated into the bodies of invertebrates, vertebrates and micoroorganisms that live in soil. These pesticides may reach the soil when the bodies of these animals containing residues in their tissues are buried.

Concern over long term effects of pesticide residues in soil has given rise to the idea that persistence of a chemical is a measurable property representing its resistence to degradation. Qualitative descriptions based largely on degradation data have been used to describe persistence, e.g. slightly, moderately or highly persistent. On a relative scale, these terms may be of some use in classifying pesticides, but have little predictive value and do not describe the conditions leading to maximum persistence. The disappearance of a pesticide from soil may not only reflect its degradability, but can also show our inability to detect its residues by conventional methods. In recent years, the use of radiolabeled pesticides has made it possible to obtain a 'mass balance' and to account for the fate of pesticides in soil.

5.1. PERSISTENCE

The word 'persistence' originates from the Latin word 'persistere' meaning 'remaining, staying in existence'. The term was employed for pesticides that retain their biological activity for a much longer period than orginally intended. For the purpose of this book, we may interpret persistence as the residence time of a pesticide in the soil environment. The residence time may be considered as the period in which the pesticide remains in soil, regardless of the method by which it is quantified. It may be expressed in units of time. Indeed, this interpretation of persistence is concerned only with the chemical and physical properties and the immediate environment of the pesticide, i.e. soil. It should be realized that the consequences of persistence can be important depending on the toxicity of the pesticide and its bioavailability.

The concept of half life is widely used in discussions of persistence of pesticides in soil. The term has been used loosely in the sence of the time required for one half of the pesticide to disappear. Hamaker (1972) pointed out that such use of the term results in ambiguity as half life also has a special meaning with respect to first order kinetics, i.e. it is essentially a rate constant. The term half life has the characteristics of (1) being a constant inversely proportional to the rate constant, (2) being independent of the concentration, and (3) representing the property that a constant percentage is lost per unit time. Thus, the time required for 25% or 90% loss is constant, just as the time required for 50% to disappear regardless of the concentration (Hamaker, 1972). For other rate laws, however, the half life is not simply related to the rate constants nor is it independent of the concentration. Hamaker (1972) suggested that the 50% loss times are not to be confused with half life in the usual sense, since they will generally depend on the concentration. The practical indices such as 50% disappearance time (DT_{50}) or 90% disappearance time (DT_{90}) have been found useful to give an idea of persistence at a given concentration. Thus, as points on the disappearance curve, the DT_{50} and DT_{90} are taken as relative disappearance times, and allow compounds to be compared for their longevity. However, it should not be used for prediction or extrapolation. Thus, DT_{90} should not and will not be given for a reaction that has only been followed to 50% disappearance (Hamaker, 1972).

Soil constitutes a major environmental sink for many pesticides from which they are taken up by plants, move into the bodies of invertebrates, pass into water or air, and are broken down. The persistence of a pesticide in soil is dependent on a host of conditions, such as soil type, organic matter content, clay content, pH, the nature of soil colloids, the microflora and microfauna present in soil, liquid and air flow through the soil, the cultural practices, and the exposure to wind, sunlight, rain and temperature, etc. Superimposed on all these factors is the chemical nature of the pesticide. Most of these conditions and factors are often interrelated and have been discussed in earlier chapters.

The concise overview (Kearney et al., 1969) of soil persistence of major classes of insecticides and herbicides is summarized in Fig. 5.1. Data for this figure were developed from a review of approximately eighty sources concerned with pesticide persistence in soils. The persistence values represent the time required for the bioactivity to reach a level of 75 to 100% of the control, or for a 75 to 100% loss of a pesticide. In addition, the values

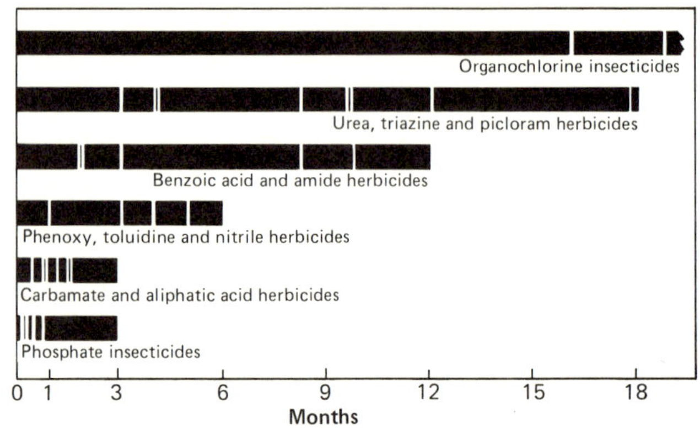

Fig. 5.1. Persistence of pesticides in soils. Reproduced from 'Chemical Fallout', 1969, p. 55, by permission of Charles C. Thomas, Publisher, Springfield, Illinois.

shown are those resulting from normal rates of application and normal agricultural conditions. Each bar represents one or more classes of pesticides. The open spaces represent the persistence of individual members within the class. These data demonstrate that the most persistent pesticides are the chlorinated hydrocarbon insecticides. The herbicides show a wide spectrum of persistence ranging from a few weeks for the carbamates and aliphatic acids to a year and a half for certain of the *s*-triazines. The organophosphorus insecticides are short lived in soils and are dissipated within a few weeks.

Kearney et al. (1969) presented 'disappearance curves' showing the loss of pesticides from soil following one or periodic applications (Fig. 5.2). They suggested that the loss of most pesticides from soil usually follows a first order reaction (Fig. 5.2a). The loss may result due to a combination of mechanisms, such as chemical alteration at clay and organic surfaces, volatilization, photodecomposition, leaching, dilution, mechanical removal and uptake, all acting on soil residues simultaneously with microbiological degradation at any given time. Once the rate constant for pesticide loss is determined, the maximum and minimum residue levels of that pesticide following periodic applications can be calculated. The periodic application of degradable pesticides would yield the type of curves shown in Fig. 5.2b. Maximum and

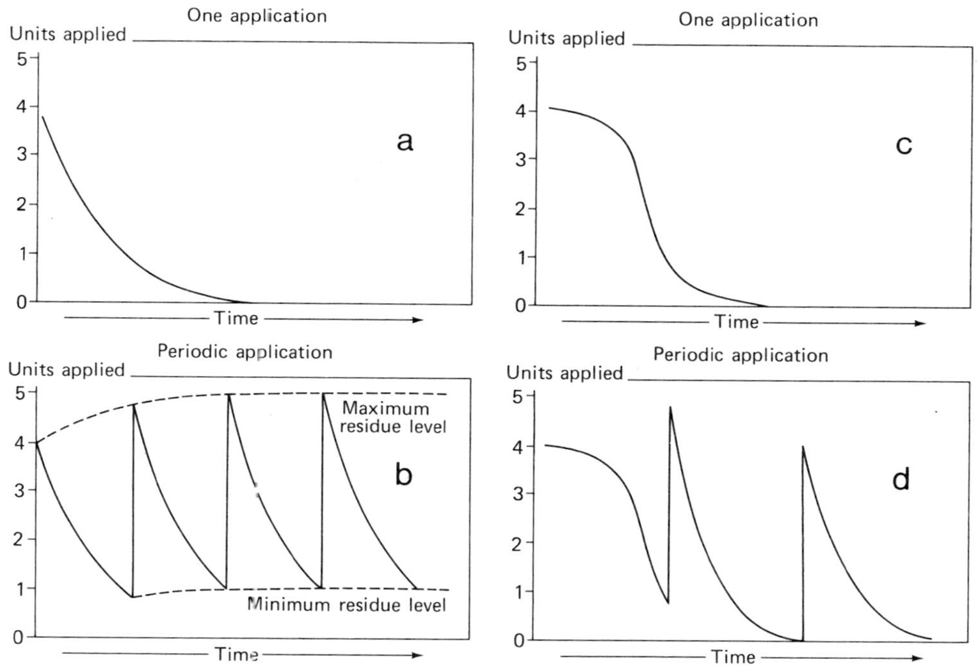

Fig. 5.2. Loss of pesticides from soil following (a,c) one, or (b,d) periodic applications. Reproduced from 'Chemical Fallout', 1969, p. 60, by permission of Charles C. Thomas, Publisher, Springfield, Illinois.

minimum residue levels would remain parallel to the base line or would not exhibit any progressive accumulation under actual field conditions, when the number of pesticide units lost in a given time equals the number of units applied. Fig. 5.2b represents a situation when 4 units of pesticide are applied periodically and it is assumed that 20% residue remains at the next application. The minimum residue level remains constant at 1 unit after the fourth application. If the pesticides were degraded to only about 50% before the next application, the minimum residue would reach to 4 units after the eleventh application. Similar curves have been developed by Hill et al. (1955), Sheets and Harris (1965), and Hamaker (1966).

When microbiological metabolism is the primary route of disappearance, a different type of loss has been observed (Fig. 5.2c). A lag phase occurs after the application in which relatively little pesticide is lost and is then followed by a rapid disappearance caused by soil microbial metabolism (Kearney et al., 1969). The pesticide applied subsequently is then rapidly degraded without lag periods and the minimum residue level remains near zero (Fig. 5.2d).

The following sections summarize the information on the persistence of the individual classes of pesticides in soil. For further details, the reader is directed to reviews by Sheets and Harris (1965); Upchurch (1966); Lichtenstein (1966); Goring (1967); Caro (1969); Kearney et al. (1969); Edwards (1966, 1976), Helling et al. (1971); and Hiltbold (1974).

5.1.1. Herbicides

Most of the organic herbicides do not build up their residues from one year to the next at the dosage levels used on agricultural crops. However, these chemicals exhibit a wide range in persistence (Fig. 5.3). A large difference may exist between herbicides within a particular class. For example, prometryn persist for only three months while propazine may persist for 18 months. A similar difference can be noted between two benzoic acid herbicides, dicamba and 2,3,6-TBA.

The *s*-triazines herbicides exhibit a varying degree of persistence in soils (Sheets, 1970; Esser et al., 1975). Methoxy *s*-triazines are usually much more persistent than chloro or

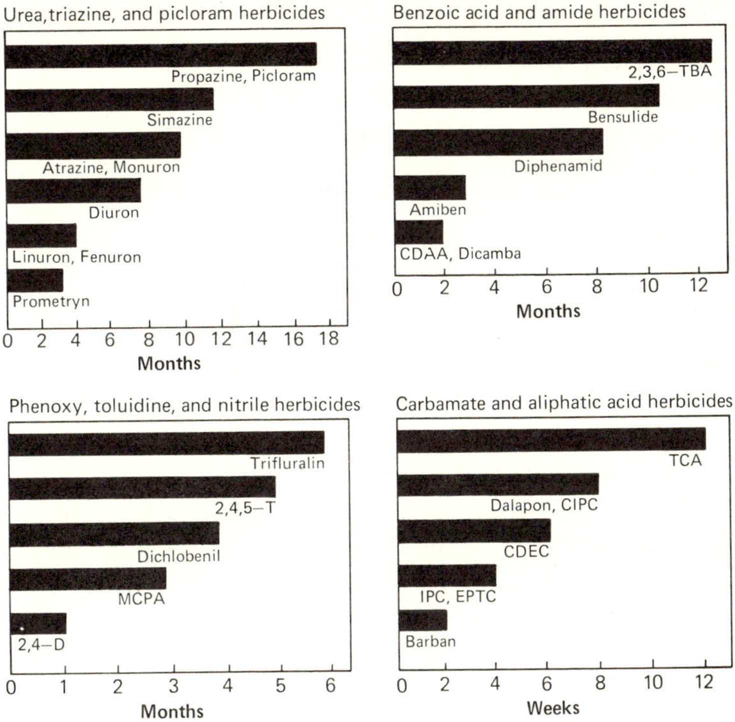

Fig. 5.3. Persistence of herbicides in soils. Reproduced from 'Chemical Fallout', 1969. p. 56, by permission of Charles C. Thomas, Publisher, Springfield, Illinois.

methylthio *s*-triazines. Sheets et al. (1962) compared the persistence of a group of *s*-triazines and observed that simetone was more persistent than any of the eight chloro substituted *s*-triazines. Some *s*-triazine herbicides, such as atrazine, simazine and propazine may persist in a soil for a year or more (Kearney et al., 1969; Sheets, 1967). Burnside et al. (1971) observed that atrazine residues increased with successive application over three years on several loam soils. Khan and Marriage (1977) reported that residues of atrazine and some of its metabolites persisted in a peach orchard soil for several years following nine consecutive annual applications of the herbicide at 4 lb/acre. A decrease in atrazine residues may result in the formation of various metabolites some of which may persist over a number of years (Marriage et al., 1975;

Khan and Marriage, 1977). Muir and Baker (1978) observed that *N*-deethylated atrazine (183) persisted at relatively high levels 12 months after the application of atrazine. The *N*-dealkylation process is considered to be the most important pathway associated with the persistence and herbicidal activity of atrazine in soils (Sirons et al., 1973). Khan and Marriage (1979) observed that simazine and the metabolite hydroxysimazine persisted for 40 and 28 months in the soil of two orchards. Furthermore, the residue levels of hydroxysimazine were at least 40 times those of simazine, 40 months and 28 months after the final application of the herbicide in the two orchards, respectively.

There are marked differences in the degree of persistence of the various substituted ureas. In this class of herbicides, increased ring chlorination and increased *N*-alkylation appears to increase persistence (Kearney, 1966). Approximately a year is required in the field for detoxification of 1 to 2 lb/acre of diuran to the point that sensitive crops will not be injured by residues (Upchurch et al., 1969). Khan et al. (1976) investigated the persistence of diuron and its degradation product, 3,4-dichloroaniline, in an orchard soil, which had received diuron annually at the rate of 4 lb/acre for 7 years. Accumulation of residues was not observed at significant levels, although carry over of the herbicide occurred between years. The degradation rate of diuron generally followed first order kinetics and the residual levels of diuron in the soil were highly phytotoxic to oat plants during the three years after the last application.

Phenoxy herbicide persistence has not been reported as sufficient to have unfavourable effects on subsequent crops. They are rapidly degraded by soil microorganisms which is in sharp contrast to the slower rates of degradation of the urea and *s*-triazine herbicides. The rate of degradation of phenoxy herbicides in soil is affected by soil moisture content and soil temperature. For example, the half lives of 2,4,5-T varies from four days at 35°C and 34% soil moisture to 60 days at 10°C and 20% soil moisture (Walker and Smith, 1979). The residual life of the phenoxy herbicides is increased if chlorine is present as a single ring substituent in the *ortho* or especially in the *meta* position. Similarly, a methyl substituent in the *ortho* position causes a longer residual life than a chloro group. The aromatic nucleus is more stable when it contains a halogen in a position *meta* to the phenolic hydroxy (Alexander and Aleem, 1961). Furthermore, cleavage of the side chain is rapid for acetate and caproate but not for propionate and valerate.

None of the carbanilate herbicides appear to persist in the soil when applied at practical rates (Kaufman, 1967). At rates of 5 to 15 lb/acre, the phytotoxic effects of chlorinated aliphatic acids, such as TCA and dalapon, dissipated within weeks (Corbin and Upchurch, 1967).

The bipyridylium herbicides, paraquat and diquat, exhibit their herbicidal action immediately upon application and, therefore, present no residual problem (O'Toole, 1966). However, their persistence will depend on clay minerals and organic matter contents of the soil. Khan et al. (1976) observed that 83 to 86% of the initial amounts of paraquat remained in an organic soil (organic matter 80%) 4 months after its application at rates 1 to 2 lb/acre. Furthermore, about 50% of paraquat was recovered from the treated organic soil 15 months after application. In a mineral soil (organic matter 1.8%) about 8% of the total paraquat applied accumulated as a residue over the 9 years period when applied annually at 2 lb/acre (Khan et al., 1975). The residue levels were below those that might be phytotoxic to oat plants.

The herbicides dicamba, 2,3,6-TBA and fenac may persist in soil for a considerably longer period (Phillips, 1968; Sheets et al., 1968).

The residual problems with substituted 2,6-dinitroaniline herbicides have been minimal (Wiese et al. 1969). About 90% of trifluralin applied to the field is rendered nonphytotoxic within about five months (Probst et al., 1967; Parks and Tepe, 1969; Duseja and Holmes, 1978). Half lives of some dinitroaniline herbicides in moist soil range from 29 to 124 days in soil under greenhouse conditions (Savage, 1978).

The uracil herbicides, such as terbacil and bromacil, persist in the soil for more than 1 year (Gardiner et al., 1969). Terbacil residues from two or three annual applications of 1.5 lb/acre or single applications of 5 to 10 lb/acre were toxic to sensitive crops for two years or more after treatment ceased (Waters and Burgis, 1968; Swan, 1972). However, in a recent study, Marriage et al. (1977) observed that terbacil residues did not accumulate in peach orchard soil after seven consecutive annual spring applications of the herbicide at the rate of 4 lb/acre.

5.1.2. Insecticides

Most of the organophosphate insecticides rarely persist into a second year. The soil type influences the persistence of organophosphorus pesticides. The disappearance rate of disulfoton and phorate in a loamy sand in winter was greater than in a silt loam in summer (Menzer et al., 1970). The organophosphates are generally short residual or decompose rapidly losing insecticidal activity within 2 to 4 weeks. Such insecticides include mecarbam, parathion, parathion methyl, phorate, mevinphos, malathion, chloropyrifos, disulfoton, dimethoate, dichlorvos, diazinon, crotoxyphos, bromophos and azinphosmethyl. The most persistent are carbophenothion with a half life of about 24 weeks (Spencer, 1968), chlorfenoinphos with one of about 12 weeks (Beynon et al., 1967) and dyfonate, which may persist for about 10 weeks (Hadaway and Barlow 1964). Trichloronate, mocap and fensulfothion are moderately persistent in soils. Chlorfenvinphos, phosfolan, dichlorfenthion and oxydisulfoton may persist over 36 weeks and are useful for soil insect control. Some of the compounds, such as phorate, may disappear rapidly, but its sulfoxide and sulfone derivatives may persist for more than 4 months in soil (Getzin and Shanks, 1970).

The insecticide phorate is more persistent in flooded, anaerobic soils than in nonflooded soils (Walter-Echols and Lichtenstein, 1978). Williams (1975) observed that in a peaty soil the insecticide chlorfenvinphos degraded very slowly, whereas in a sandy soil persistence was much shorter. Dyfonate is moderately persistent, about 5% of the insecticide remains in mineral soil, more than two years after application (Saha et al., 1974). Khan et al. (1976) also observed that dyfonate was moderately persistent in organic soil. The formulation also affects the organophosphates persistency. For example, 50% of azinphosmethyl sprayed as an emulsion on a soil surface disappeared within 12 days, whereas that in granules dissipated 50% within 28 days (Lichtenstein, 1972).

The carbamate insecticides are slightly to moderately persistent in soils. Almost all carbamate insecticides have half lives of short duration in soils, ranging from only a few days to a few weeks. The parent compound and known toxic metabolites virtually undergo destruction in 1 to 4 months. The systemic methylcarbamates, such as carbofuran, could be an exception, since the reported half lives range from 18 to 378 days (Tsukamoto and Suzuki, 1964). The carbamate insecticides are rarely used against pests in soil and pose very little persistence problem.

The organochlorine insecticides are much more persistent than other pesticides. The most common residues in soils are of DDT and related compounds. The decay of DDT residues in forest soils is very slow (Owen et al., 1977), and may well approximate the 35 year half life suggested by Dimond et al. (1970). Owen et al. (1977) cautioned about the implication of their findings as any additional applications of DDT will be additive in the foreseeable future. Harris et al. (1977) reported DDT residues in soils from fifteen farms being highest in orchard > vegetable > tobacco > field crop soils. Cyclodiene insecticide residues were present in soil on thirteen of the 15 farms being highest in vegetable > tobacco > field crop > orchard soils. The soils were sampled in 1974 as part of a long term study initiated 10 years earlier (Harris et al., 1977). Saha et al. (1968) reported more than 0.1 ppm dieldrin residues in soils from 16 fields. Heptachlor, heptachlor epoxide, endrin, and aldrin were also present in soil from ten fields. Saha and Sumner (1971) observed that all but 2 of the 41 agricultural soil samples from 21 vegetable farms had more than 0.01 ppm of total organochlorine insecticide residues, with a maximum of 6.9 ppm. Table 5.1 shows half lives of some organochlorine insecticides (Menzie, 1972). These values were obtained when the insecticides were worked into the soil. If the approximate half life of an insecticide is known, predictions can be made regarding the likelihood of its accumulation in soil after successive treatments (Hamaker, 1966). Therefore, for an exponential breakdown of insecticides, half lives up to one year will result in residues of not more than twice the annual addition whether this is divided or added all at once. The maximum accumulation will be about six times the annual application of insecticide for a half life of four years; rising to not more than 15 times the annual treatment for a half life of 10 years (Hamaker, 1966). Decker et al. (1965) demonstrated the usefulness of such calculations. They calculated the expected amounts of residues of aldrin in 35 corn field in Illinois after regular annual treatments for

Table 5.1

Half lives of some organochlorine insecticide in soils (Menzie, 1972)

Insecticide	Approximate half life (years)
DDT	3-10
Heptachlor	7-12
Isodrin/endrin	4-8
Toxaphene	10
Aldrin	1-4
Dieldrin	1-7
Chlordane	2-4
BHC	2

up to 10 years, then sampled the fields and analyzed the residues. The agreement between the predicted and actual residues was remarkably close.

Edwards (1966) summarized the relative persistence of the various organochlorine insecticides in soil (Table 5.2). On the average, DDT persists longest in soil, followed by dieldrin, endrin, lindane, chlordane, heptachlor, and aldrin in order of decreasing persistence. Edwards (1966) also drew regressions based on the available data in the literature (Fig. 5.4). It can be seen that the disappearance of DDT approximates to a simple

Table 5.2

Persistence of some organochlorine insecticides in soil (Edwards, 1966)

Insecticide	Average dosage active ingredient (lb/acre)	Time for 95% disappearance (years)
DDT	1-2.5	4-30
Heptachlor	1-3	3-5
Aldrin	1-3	1-6
Chlordane	1-2	3-5
Dieldrin	1-3	5-25
Lindane	1-2.5	3-10
Telodrin	0.25-1	2-7

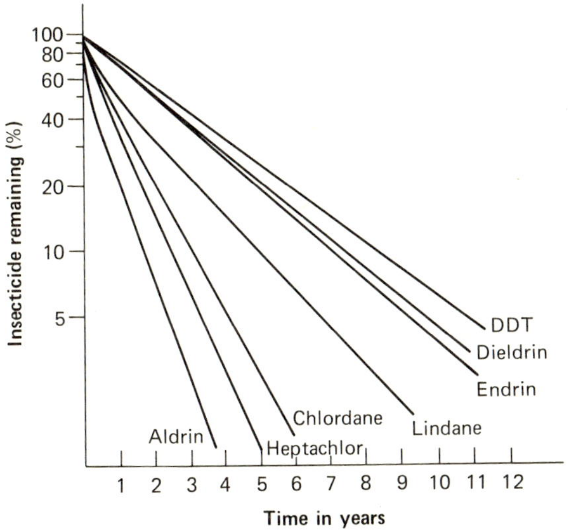

Fig. 5.4. Breakdown of organochlorine insecticides in soil. Reproduced from 'Persistent Pesticides in the Environment' 1976, p. 16, by permission of the Chemical Rubber Co., CRC Press, Inc.

exponential curve. At the other extreme, aldrin differs considerably, and this has been attributed to greater volatility. Other organochlorine insecticides such as hexachlorobenzene are persistent in soil for many years (Isensee et al., 1976) and γ-chlordane is more stable in soils than its α-isomer (Tafuri et al. 1977).

The widespread use of organic soils for vegetable crop production requires the effective use of insecticides for pest control. In general, high insecticide residue levels have been found in agricultural organic soils. Recently Miles and Harris (1978b) summarized results on the occurrence of insecticide residues in organic soils on 28 farms located in six widely separated vegetable growing areas in southwestern Ontario, Canada, (Table 5.3). Organochhlorine insecticide residues were detected in all soils. Organophosphorus insecticide residues were present in soil on 26 farms with ethion predominating. Carbamate insecticides, mainly carbofuran, were found on ten farms, with one soil containing 8.7 ppm total carbofuran. Once incorporated into organic soil, the insecticides may persist (Miles et al., 1978) but are not adsorbed from soil by

TABLE 5.3

Insecticide residues (ppm) detected in agricultural organic soils (Miles and Harris, 1978)

Organochlorine		Organophosphorus		Carbamate	
Compound	Residue	Compound	Residue	Compound	Residue
Total DDT	T[1]-28.8	Ethion	T-7.81	Carbofuran	T-7.33
Aldrin	T - 0.06	Fonofos	0.06-1.10	3-Keto carfuran	T-1.30
Dieldrin	0.02-1.74	Dichlofenthion	T-0.31	Carbaryl	0.03-0.08
Endrin	T - 0.86	Leptophos	0.03-0.30		
Endosulfan	0.03-1.79	Diazinon	T-0.29		
γ-Chlordane	0.02-0.03	Parathion	0.06-2.50		
Heptachlor	0.06-0.08				

[1]T = trace (<0.1 ppm for DDT, carbofuran and 3-keto carbofuran; <0.01 ppm for cyclodienes, dichlofenthion; <0.02 ppm for ethion, diazinon)

crops to any great extent (Harris and Sans, 1969). Miles and Harris (1978a) suggested that these residues do not appear to constitute a serious environmental hazard except when they are transported into a drainage system, where some persist and others degrade according to their individual chemical and physical properties. Khan et al. (1976) observed that 4 months after treatment, about 40 to 48% of the initial amounts of fonofos remained in an organic soil. However, 16½ months after treatment, the amount of fonofos present was only 16 to 26% of the insecticide initially recovered from the soil.

Residues of toxaphene may persist in soil for several years. In crop land under regular use, a recovery rate in the order of 10 to 30% has been observed, 1 to 3 years after the last application (Stevens et al., 1970). Nash and Woolson (1967) determined pesticide residue levels in soil 14 years after the soil had been treated. Of the nine pesticides tested, toxaphene was the most persistent. Forty five percent of the amount of toxaphene applied still remained at the end of the test period and the authors concluded that toxaphene had a half life of 11 years. Hermanson et al. (1971) tested the persistence of several insecticides over a period of 11 years. Toxaphene was the fourth most persistent of the seven organochlorine insecticides investigated. Nash et al.

(1973) determined several pesticide residues 20 years after the soils were treated. Toxaphene residues were the most persistent and represented 45% of the original application.

The pyrethroid insecticide, permethrin, degrades in soil rapidly and the half life averaged 28 days or less (Kaufman, 1977). Williams and Brown (1979) observed that the degradation of permethrin and WL43775 in five soils was rapid, resulting in half lives of approximately 3 weeks for (Z)- and (E)-permethrin and 7 weeks for WL43775. However, in another soil very little degradation occurred and the recovery after 16 weeks was greater than 75% for (Z)-permethrin and WL43775, and slightly less for (E)-permethrin. Bélanger and Hamilton (1979) reported that permethrin applied to an organic soil persisted for the initial 28 days and declined slowly during the rest of the season.

5.1.3. Fungicides

Most of the organic fungicides are biodegradable and persist in soil for a very short period. Inorganic fungicides that contain heavy metals persist longest. The copper, tin or mercury residues, formed as a result of a breakdown of the fungicide, persist in the soil for a long period. Among the organic fungicides, the most persistent are quintozene, which breaks down in several months or even in one year, benomyl and methyl thiophenate which persist from 6 months to 2 years according to soil type, and thiram which may persist for several months. Under field conditions maneb has an overall half life in soil between 4 and 8 weeks and the half life for ETU is less than one week (Rhodes, 1977).

5.1.4. Other Pesticides

Inorganic arsenicals such as arsenic trioxide, sodium arsenite, and calcium arsenate have been used for many years as soil sterilants and nonselective herbicides. The organic arsenical compounds include dimethylarsenic acid and methylarsenic acid, which can break down into compounds that persist in soil longer than other herbicides or their residues. Applications of arsenical pesticides on orchards, vineyards, and tobacco crops result in an accumulation of very large amounts of arsenic. Arsenic residues in soil up to approximately 220 ppm level have been reported in orchards and vineyards (Stevens et al., 1970). The mean values of arsenic residues in some agricultural soils in Canada and U.S. ranged from 3.7 ppm to 18.2 ppm level (Miles, 1968; Steevens et al., 1972).

5.2. BOUND RESIDUES

It is a common observation that a portion of pesticide residues remain in soil after solvent extraction. This has been shown by using radiolabeled pesticides. The use of combustion technique with the extracted soils has made it possible to release and detect the radiolabeled unextractable or bound residues in the form of $^{14}CO_2$.

The soil bound residue has been defined as "that unextractable and chemically unidentifiable pesticide residue remaining in FA, HA and humin fractions after exhaustive sequential extraction with nonpolar organic and polar solvents" (U.S. Environmental Protection Agency, 1975; Kaufman, 1976). The bound residues may be considered as hidden residues that keep an intact molecule capable of subsequent release and exertion of long term biological effects. On the other hand, it is possible that binding of soil residues may represent the most effective and safest method of decontamination by rendering the molecule innocuous and allowing slow degradation in the bound state to products that pose no short or long term problems (Kearney, 1976).

In the formation of bound residues with the herbicide propanil, the bulk of the immobilized aromatic propanil moiety is chemically bound to HA to form a humus-3,4-dichloroaniline complex (Bartha, 1971). More than half of 3,4-dichloroaniline is converted to non-hydrolyzable residues, which may be integrated into the soil organic matter nuclei (Hsu and Bartha, 1973). A 190 day laboratory experiment with radiolabeled 3,4-dichloroaniline demonstrated that the hydrolyzable residues declined with time, whereas the nonhydrolyzable residues did not, or did so at a much slower rate (Hsu and Bartha, 1976). Viswanathan et al. (1978) detected about 90% of bound ^{14}C in soil, 1.5 years after soil treatment with radiolabeled 3,4-dichloroaniline. Soil bound residues have also been reported for the herbicide propanil (Chisaka and Kearney, 1970), the fungicide 2,6-dichloro-4-nitroaniline (Van Alfen and

Kosuge, 1976), insecticides fonofos (Flashinski and Lichtenstein, 1974) and carbaryl (Kazano et al., 1972). The insecticide phosalone is degraded rapidly in both moist and flooded soil with an accumulation of ^{14}C from the benzoxazolone moiety into the soil bound residue fraction (Ambrosi et al., 1977a). The ^{14}C in the bound fraction is most extensively associated with the FA fraction, where it appears to be fairly stable. Ambrosi et al. (1977b) reported that the herbicide oxadiazon applied to a soil had bound residues of up to 13.3% after 25 weeks. Furthermore, the distribution of ^{14}C in the bound residue fraction of the moist soil was FA > HA or humin, whereas ^{14}C was fairly evenly distributed in the flooded bound residue fraction.

Katan et al. (1976) found that the total radiocarbon (extractable and bound) recovered 28 days after treatment of a loam soil with ^{14}C-parathion still amounted to 80% of the applied dose. Of this, 35% was extractable and associated with parathion and 45% was bound (Fig. 5.5). Binding of ^{14}C-residues was related to

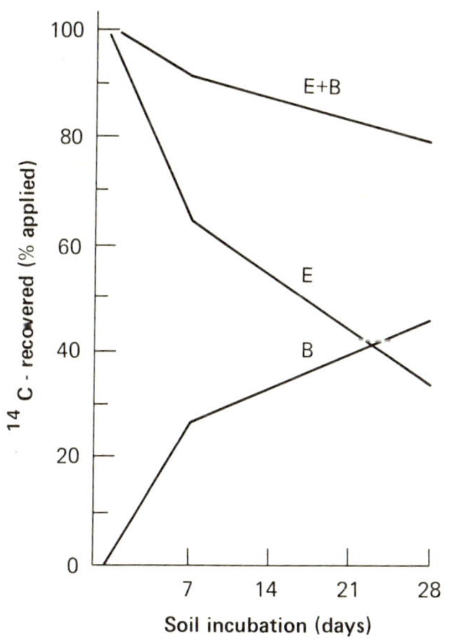

Fig. 5.5. Binding and extractability of [phenyl-^{14}C] parathion in soil. Curve B, bound parathion; curbe E, extracted parathion; and curve E + B, the total (Katan et al., 1976).

the activity of soil microorganisms. In a subsequent study Lichtenstein et al. (1977) investigated the extractability and formation of bound ^{14}C-residues in a loam soil with nonpersistent insecticides, ^{14}C-methyl parathion and ^{14}C-fonofos, and with the persistent insecticides, ^{14}C-dieldrin and *p,p*-^{14}C DDT (Fig. 5.6). It was observed that ^{14}C-methyl parathion was rapidly bound to loam soil, where up to 41% of the applied ^{14}C-insecticide residues could not be extracted after a 7 day incubation period.

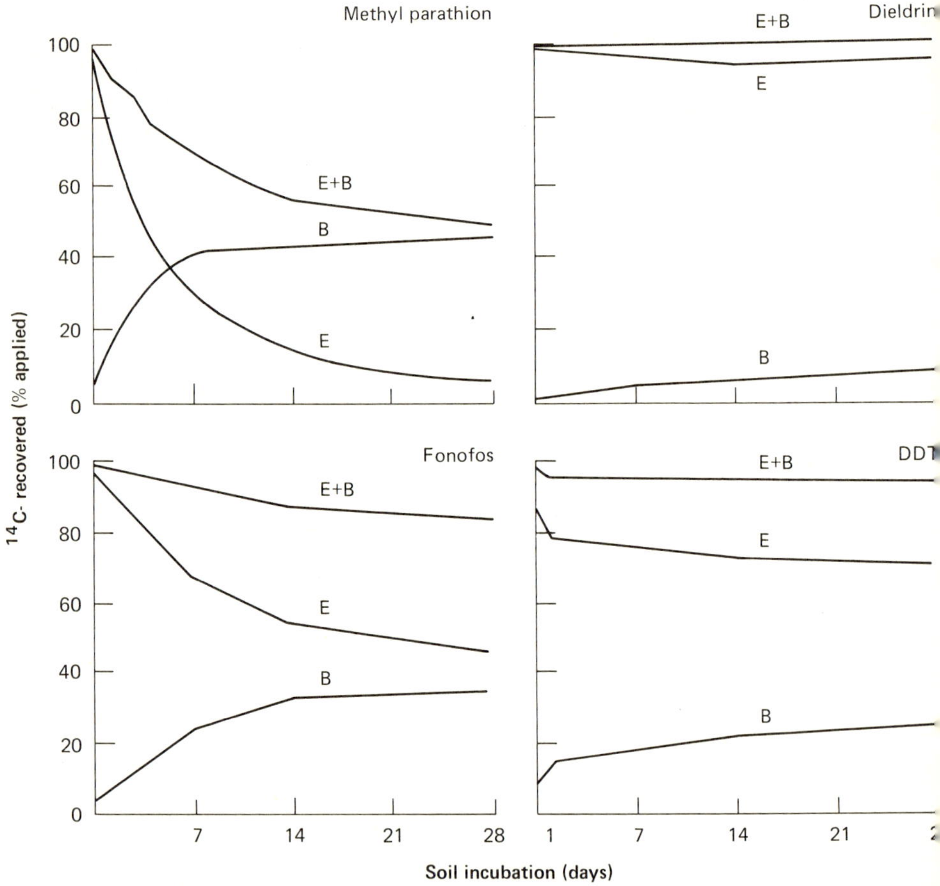

Fig. 5.6. Binding and extractability of ^{14}C-labeled insecticides in soil. Curve B, bound insecticide; curve E, extracted insecticide; and curve E + B, the total (Lichtenstein et al., 1977).

Furthermore, only 7% of the applied radiocarbon was extractable 28 days after soil treatment, whereas ^{14}C-bound residues amounted to 43% of the applied dose. With ^{14}C-fonofos, however, 28 days after soil treatment, 47% of the applied dose was extractable and 35% of the applied radiocarbon was bound (Fig. 5.6). With the persistent insecticides, dieldrin and DDT, smaller amounts of bound residues were formed. Thus, they differed from the organophosphorous compounds in their relative low binding properties and their high extractability from soils. Lichtenstein et al. (1977) also observed that only a fraction of the radiocarbon extracted from ^{14}C-methyl parathion treated soil was associated with the parent compound, whereas extractable ^{14}C-residues from the other insecticide treated soils were primarily due to the presence of the parent compounds. Contrary to the results obtained with ^{14}C-parathion (Katan et al., 1976), the binding of ^{14}C-fonofos in soil was not related to the presence of microorganisms (Lichtenstein et al., 1977). The role of microorganisms in soil binding phenomena consists of degrading ^{14}C-parathion to compounds that are more tightly bound to soil than the parent insecticide (Katan and Lichtenstein, 1977).

In a recent study Wheeler et al. (1979) investigated ^{14}C-trifluralin binding to two soils. The percentage of bound ^{14}C increased with time, the silty clay loam soil (organic carbon 3.9%) bound a higher percentage of ^{14}C than did a sandy loam organic carbon 0.9%). In accordance with the earlier findings of Katan and Lichtenstein (1977) with the amino analogue of parathion, Wheeler et al. (1979) also observed a significant relationship between the amount of binding and the substitution on the amino nitrogen. Recently, Spillner et al. (1979) implicated 2-methylhydroquinone, an oxidative product of 3-methyl-4-nitrophenol, as the precursor to the formation of fenitrothion bound residues in aerobic soils. However, under anaerobic conditions binding will proceed through the amino intermediates. Golab et al. (1979) investigated the degradation of trifluralin in a field soil. After three years, 38% of the applied trifluralin remained in soil as a bound residue. α,α,α-Trifluorotoluene-3,4,5-triamine (174), a degradation product of trifluralin, appeared to be a key compound in the formation of soil bound residues.

Recently Helling and Krivonak (1978a) observed that soil bound residues of six [phenyl-^{14}C] dinitroaniline herbicides constituted 7-21% of the original ^{14}C added to aerobically incubated silt loam soil. Bound residues of butralin from another silt loam soil were 3 and 13% after aerobic and anaerobic incubation, respectively. Helling and Krivonak (1978a) summarized the work of several other investigators dealing with the bound dinitroaniline content in aerobic soils. Their results for trifluralin were similar to those reported earlier but data on dinitramine and fluchloralin differed substantially (Table 5.4). The anaerobic bound residues

TABLE 5.4

Bound residue levels of some pesticides in soils

Pesticide	Time[1]	Bound residues (% of applied)	Reference
Herbicides			
Propanil	20 d	54-73	Bartha (1971)
Benefin	12 m	14	Golab et al. (1970)
Dinitramine	8 m	55	Smith et al. (1973)
"	5 m	8	Helling and Krivonak (1978)
Trifluralin	12 m	8	Probst et al. (1967)
"	12 m	50	Golab and Amundson (1975)
"	7 m	7	Helling and Krivonak (1978)
"	36 m	38	Golab et al. (1979)
"	63 d	72	Wheeler et al. (1979)
Fluchloralin	5 m	10	Otto and Dresher (1973)
"	7 m	21	Helling and Krivonak (1978)
Profluralin	7 m	11	Helling and Krivonak (1978)
Chlornidine	7 m	17	Helling and Krivonak (1978)
Butralin	7 m	17	Helling and Krivonak (1978)
Oryzalin	6-36 m	30-35	Golab et al. (1975)
"	12 m	56	Golab and Amundson (1975)
Isopropalin	12 m	15-27	Golab and Althaus (1975)
"	12 m	20	Golab and Amundson (1975)
Prometryn	5 m	45	Khan and Hamilton (1980)
Oxadiazon	25 w	13.3	Ambrosi et al. (1977b)
3,4-Dichloroaniline	18 m	90	Viswanathan et al. (1978)
Insecticides			
Parathion	28 d	45	Katan et al. (1976)
Methylparathion	28 d	43	Lichtenstein et al. (1977)
Fonofos	28 d	35	Lichtenstein et al. (1977)
Phosalone	84 d	80	Ambrosi et al. (1977a)
Fenitrothion	50 d	48-50	Spillner et al. (1979)
Pirimicarb	12 m	20-60	Hill (1975)
Dieldrin	28 d	6.5	Lichtenstein et al. (1977)
p,p-DDT	28 d	25	Lichtenstein et al. (1977)

[1]d = days, w = weeks, m = months

were associated with more 'humified' organic matter fractions than were residues formed during aerobic incubation. However, it was noted that distribution of ^{14}C was very broad in soil components and that the classical FA/HA fraction distorted the picture. Based on thermoanalytical investigations it was postulated that the parent herbicide must be chemically bound to soil to produce bound residue and it is very unlikely that bound ^{14}C had become a part of a highly condensed nucleus of soil organic matter (Helling and Krivonak, 1978a). Spillner et al. (1979) observed that bound radiocarbon [(ring-^{14}C) fenitrothion] was associated mainly with HA and FA fractions. The binding was explained through an intermediate, 2-methylhydroquinone, which copolymerizes with humic substances during their formation to yield radioactive products incorporated into the soil organic matter.

The analytical methods employed in studies described above involved combustion of the soil to release $^{14}CO_2$ for quantitating nonextractable or bound ^{14}C residues. However, this technique results in the destruction of bound residues identity. Recently, a novel technique was developed in the author's laboratory to determine and chemically identify the bound residues of the herbicide prometryn in the field treated and laboratory incubated soil samples. The technique involves high temperature distillation to release bound residues (Khan and Hamilton, 1980). The equipment used is shown in Fig. 5.7. An air dried soil sample containing bound residues was placed into a porcelain boat and inserted into the middle of the quartz tube. One end of the tube was closed and the other end was connected with a series of traps (Fig. 5.7). The furnace was heated from room temperature to 800°C (@ 15°C/minutes) and maintained at this temperature for about 15 minutes. Helium was used as a sweep gas at a flow rate of about 50 ml per minute. At the end of the experiment, the collection U-tube (Trap II), as well as the quartz tube, was thoroughly washed with methanol and the material in different traps was then processed as described in Fig. 5.8. Soil samples containing ^{14}C-residues were also combusted in a Packard sample oxidizer to produce $^{14}CO_2$.

The amounts of the extractable ^{14}C-residues recovered from soil decreased over an incubation period of 150 days (Fig. 5.9). This in turn, corresponded to an increase in the formation of soil bound ^{14}C-residues. Thus, by the end of the incubation period, extractable ^{14}C-residues decreased to 36.5% while the bound

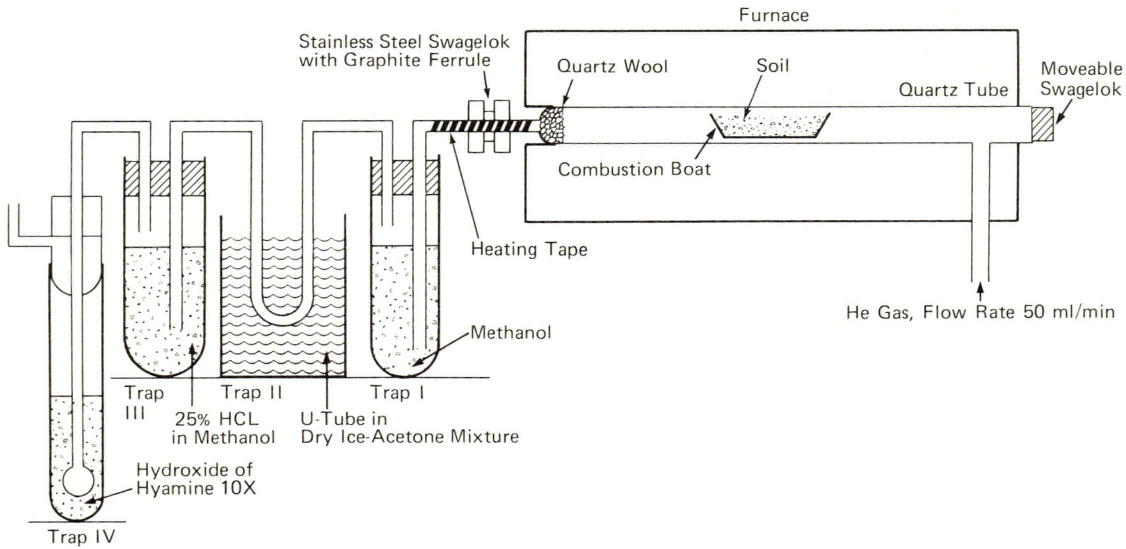

Fig. 5.7. Apparatus for high temperature distillation of soil samples (Khan and Hamilton, 1980).

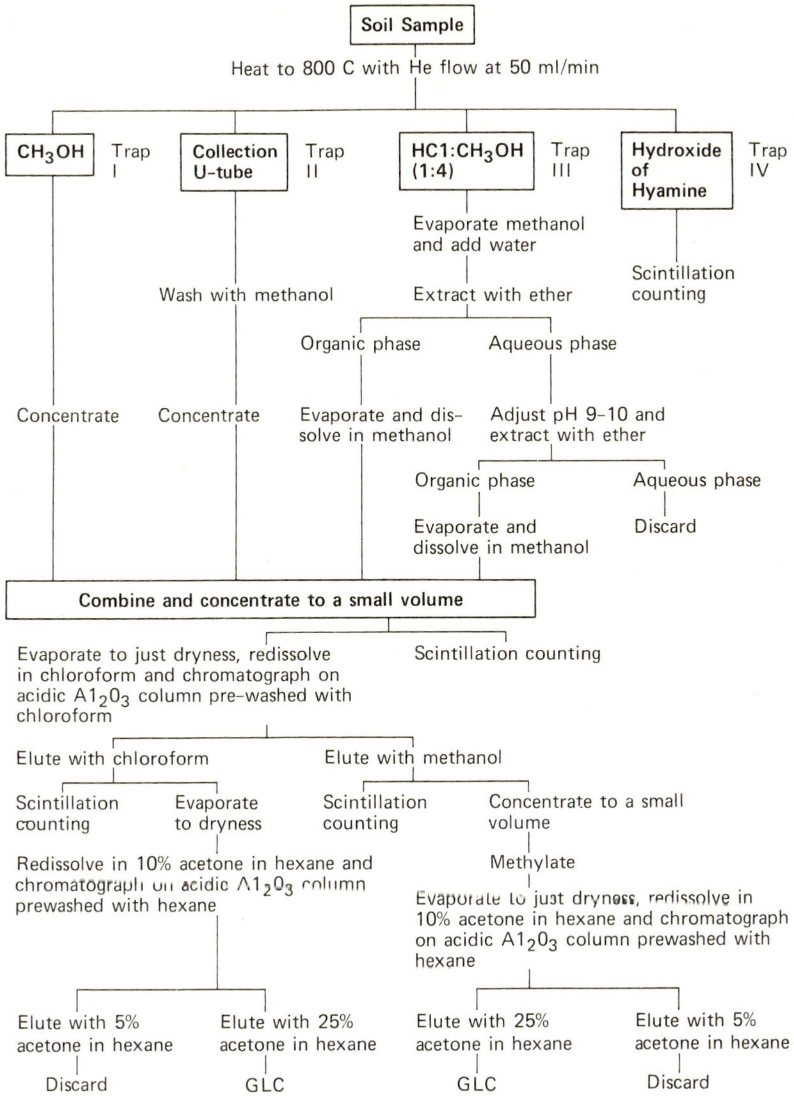

Fig. 5.8. Schematic diagram for the analysis of bound residues (Khan and Hamilton, 1980).

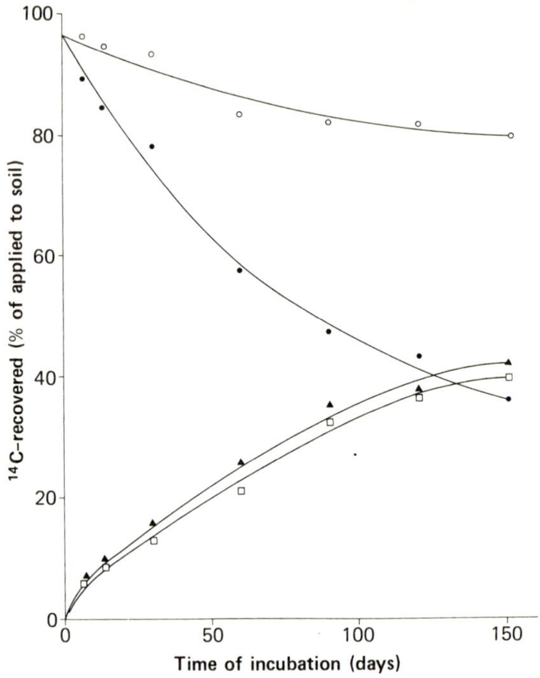

Fig. 5.9. Extractable and bound residues of ^{14}C-prometryn in an organic soil during a 150 day incubation period. •=extracted ^{14}C; ▲=bound ^{14}C determined by combusting the soil to release $^{14}CO_2$; □=bound ^{14}C determined by high temperature distillation; and ○=extractable plus bound (Khan and Hamilton, 1980).

^{14}C-residues (determined by combustion to $^{14}CO_2$) increased to 43.0% of the initially added ^{14}C. The total radioactivity recovered at the end of 150 days amounted to about 80% of that initially applied. Similarly, the radioactivity recovered by high temperature distillation of samples increased with incubation time and by the end of 150 days amounted to about 40% of the initially added ^{14}C. However, the amounts of ^{14}C recovered by this technique were slightly lower than those obtained by combustion to $^{14}CO_2$ (Fig. 5.9).

The amount of radioactivity in the combined material from traps I, II and III (Fig. 5.7 and 5.8) was 73.9 to 80.9% of the total ^{14}C released by high temperature distillation. The remaining radiocarbon was thermally decomposed to $^{14}CO_2$ (trap IV). Analysis of the combined material (traps I, II and III) according to the scheme depicted in Fig. 5.8 revealed that over an incubation period of 150 days the radioactivity of the chloroform soluble material decreased from 88.8 to 62.1% of the total in the three traps. A corresponding increase in radioactivity from 16.9 to 25.9% was observed in the methanol soluble material. Gas chromatographic-mass spectrometric analyses indicated that the chloroform eluate contained mainly prometryn.

The application of high temperature distillation technique to the field treated soil samples made it possible to release the bound prometryn residues. The latter were collected in suitable solvents, purified and analyzed by gas chromatography and gas chromatography-mass spectrometry (Khan and Hamilton, 1980). The bound prometryn residues in the field samples will not be detected in the routine residue analysis involving exhaustive solvent extraction of soil samples. This would result in an underestimation of the total prometryn residues in soil. It was observed that 64 days after treatment with the herbicide at 2 and 4 lb/acre, the total prometryn concentration determined in an organic soil constituted about 66 and 57% extractable and 34 and 43% bound residues, respectively (Khan and Hamilton, 1980).

Chemical identification of the bound entities have been rarely reported. Hill (1975) attempted to identify the bound radioactive residues of the insecticide pirimicarb resulting from incubation of soil under aerobic and flooded conditions. Booth et al. (1975) determined the release of soil bound fluchloralin in water by combusting soil samples and analyzing water samples. The high temperature distillation technique developed by Khan and Hamilton (1980) enables the identification and measurement of bound residues in the laboratory and field treated soils. Contrary to the general consensus that the unextractable or bound pesticide becomes an integral part of the matrix without recognizable relationship to the original compound, it appears that a considerable proportion of such residues in soil may comprise of the parent molecule. Since the bound pesticides may constitute a significant part of the total residues in soils, special attention should also be given to this form of residues in assessing the disappearance of pesticides in soil.

The question of the significance of bound residues has become an important one at the current time. It is important that we should be able to predict what effects these compounds will have on the chemical and biological systems if they should be released in the soil. Suss and Grampp (1973) reported that mustard plants could take up residues of ^{14}C-monolinuron, which could not be removed from soil with five acetone extractions. Much of the current information pertaining to the nature and the potential biological activity of the bound residues has been published by Lichtenstein and his co-workers. Katan et al. (1976) investigated the toxicity to fruit flies (*Drosophila melanogaster* Meigen) of parathion bound residues. The insects were exposed for 24 hours to sandy soil containing bound, as well as freshly added parathion. None or very few mortalities were observed with exposure to bound residue soil. However, mortalities of the insects were considerably higher when exposed to soil containing freshly added parathion. In a later study, Lichtenstein et al. (1977) reported similar results for methyl parathion and fonofos thereby indicating that bound residues are biologically less active. In a recent study, Fuhremann and Lichtenstein (1978) investigated the potential release of bound residues of methyl ^{14}C-parathion from soil in the presence of earthworms and oat plants, and the potential pickup and possible metabolism of the ^{14}C residues by these organisms. After worms had lived for 2 to 6 weeks in soil containing 32.5% bound residues of the applied insecticide or several crops of oats had grown in it, 58 to 66% and 82 to 95% soil bound ^{14}C-residues were taken up by earthworms and oat plants, respectively. the majority of soil bound residues taken up by earthworms had again become bound in the worms, whereas most of the residues in the oat plants were extractable (Fuhremann and Lichtenstein, 1978). The biological availability of bound dinitroaniline herbicides was recently investigated by Helling and Krivonak (1978b). Uptake of bound residues was found to occur from soil. However, the evaluation of uptake and bioactivity of bound herbicides was complicated by phytotoxic concentrations of Mn in the soil. In their experiment, extraction of a moderately acidic soil with benzene-methanol led to Mn toxicity in soybeans. Helling and Krivonak (1978b) postulated that any common pesticide extraction technique would also kill or inhibit soil microorganisms to the degree that plants subsequently grown in the soil might be artificially affected.

Data presented by Lichtenstein and his co-workers clearly indicate that soil bound insecticide residues are not excluded from environmental interactions (Katan et al., 1976; Lichtenstein et al., 1977, Fuhremann and Lichtenstein, 1978). The bound residues will not be detected in routine residue analysis. Thus, the disappearance of a pesticide from soil should not only be described by its degradation, volatilization or leaching but should also include the formation of unextractable or bound residues. Lichtenstein et al. (1977) stated that "the expression 'disappearance' and 'persistence' of pesticides, so widely used during the last two decades, should be reassessed to consider the bound products".

5.3. PESTICIDES IN SOIL ANIMALS

In agricultural soils, many invertebrates take up pesticides from soil into their body and may concentrate pesticides several times greater in their tissues than those in the surrounding soil. The animals that feed upon these invertebrates may in turn concentrate these residues to levels that may kill them or affect their normal activities. Residues of DDT, BHC, aldrin, dieldrin, methoxychlor, chlordane, endrin and heptachlor have been found in soil invertebrates. The subject has been reviewed by Edwards and Thompson (1973) and Edwards (1976). It has been observed that some organochlorine insecticides are metabolized in worms (Edwards and Thompson, 1973). Residues of organochlorine insecticides have also been reported in slugs and carabid beetles (Edwards and Thompson, 1973). Wheatly and Hardman (1968) plotted earthworm concentrations (wet weight) against the soil concentrations (dry weight) of the organochlorine insecticides. A linear relationship on a log-log scale was obtained (Fig. 5.10), the concentration factor being tenfold at the lowest concentration and falling to unity at the highest concentrations. Edwards and Thompson (1973) reported from all the available data the degree of concentration of organochlorine compounds from soil to slugs (Fig. 5.11). Slugs concentrate *t*-DDT and aldrin-dieldrin soil residues by an average of seven to eight times. Gish (1970) carried out studies on residues of organochlorine insecticides in earthworms, slugs, and snails inhabiting several treated fields in orchards in the United States. It was found that earthworms accumulated more than snails, and slugs more than earthworms (Table 5.5).

Uptake and accumulation of some organophosphorus and carbamate insecticides from treated soils by earthworms and slugs has also been reported (Edwards and Thompson, 1973). However, no data on

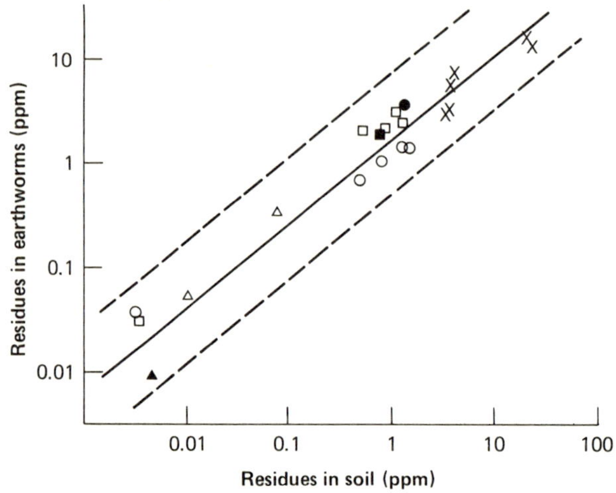

Fig. 5.10. Average organochlorine residues in earthworms plotted against soil residues (Wheatley and Hardman, 1968). Reproduced from 'Ecology of Pesticides', 1978, p. 80, by permission of John Wiley & Sons, Inc.

× DDT
■ DDE
▲△ Lindane
● Aldrin + Dieldrin
○ Dieldrin, A. *longa*
□ Dieldrin, A. *clorotica*

residues of organophosphorus insecticides in beetles have been reported.

Since invertebrates such as earthworms and molluscs could concentrate pesticides from soil into their fatty tissue (Thompson, 1973), more attention to pesticide residues in soil invertebrates is necessary in order to accurately assess the hazards caused by pesticide residues in soil animals.

5.4. PLANT UPTAKE

The presence of pesticides in soil may lead to their residues in plants grown in the contaminated soil. An excessive amount of pesticide in the soil is a necessary condition of its uptake from the soil by a plant. However, the rate of uptake may differ for pesticides that are equally persistent. Lichtenstein et al. (1965a) observed that heptachlor is absorbed more readily than

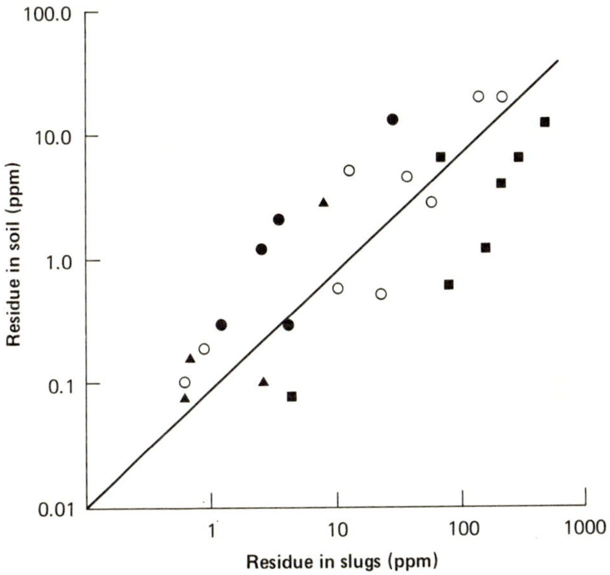

Fig. 5.11. The concentration of insecticides from soil to slugs (Edwards and Thompson, 1973). Reproduced from 'Ecology of Pesticides', 1978, p. 86, by permission of John Wiley & Sons, Inc.

■ Edwards (Unpublished data)
○ Gish (1970)
● Davis (1968)
▲ Cramp and Olney (1967)

TABLE 5.5.

Residues of organochlorine insecticides in earthworms slugs, and snails inhibiting treated fields and orchards (Gish, 1970)[1]

Insecticide	Residues (ppm, dry weight)			
	Soil	Earthworms	Slugs	Snails
DDT	0.08-5.4	1.1-54.9	10.3-36.7	0.32-0.38
DDE	0.12-4.4	1.4-17.6	4.2-15.4	0.70-1.06
DDD	0.01-5.6	0.8-18.7	2.6-14.0	0.83-1.68
Aldrin	-	0.02-0.2	0.2	-
Dieldrin	0.01-0.02	0.04-0.82	0.2-11.1	0.02-0.07
Endrin	0.01-3.5	0.4-11.0	1.1-114.9	2.72

[1]The treatment of insecticides ranged from 3 to 18 lb/acre

aldrin. If the pesticides are mixed evenly in soil, it is unlikely that the rate of transport in soil will limit their uptake by plant roots (Graham-Bryce, 1968). However, some relationship is likely to exist between residues in soil and residues in plants grown in the soil. A correspondence of pesticide residues in layers of soil and in some root crops have been observed (Lichtenstein et al., 1965a). This indicates that the pesticide passes directly from the soil to these plants. However, not all residues that pass from soil to plants are transferred in that way. Residues of aldrin and heptachlor from soil were absorbed by the cucumber plant roots and translocated through the stems to the cucumbers (Lichtenstein et al., 1965b).

Pesticides are absorbed into crops most readily from sandy soils and least readily from muck soils containing a high content of organic matter (Table 5.6). Similarly, concentration of insecticides is more effective in sandy soil than in muck soil. A significant proportion of the pesticide can be dislodged by rain soon after application. However, rain late in the season will have little effect on the firmly bound residues in soil.

TABLE 5.6

Movement of insecticides from soil into carrots (Oloffs et al., 1971)

Insecticide	Sandy soil			Muck soil		
	Source[1]	Plant[1]	Conc. or dilution factor[2]	Source[1]	Plant[1]	conc. or dilution factor[2]
BHC	0.095	0.0249	0.262	0.693	0.0225	0.032
Heptachlor	0.066	0.0063	0.095	4.563	0.017	0.003
Heptachlor epoxide	0.375	0.033	0.088	3.563	0.0215	0.006
Dieldrin	1.165	0.0455	0.039	8.563	0.0251	0.003
p,p-DDT	4.650	0.0374	0.008	10.217	0.0265	0.003

[1]ppm (Dry weight for soil, fresh weight for plants)
[2]Concentration or dilution factor = $\frac{\text{amount in plant}}{\text{amount in soil}}$

In general, the nonpolar pesticide tends to be absorbed by the root surface, whereas the polar compound readily passes through the epidermis and is translocated through the plant. Accumulation of a pesticide in a plant usually is dependent upon the concentration of the residues in the soil. The total amount of the pesticide in plant may increase with time if the compound is long lived. Water solubility of the pesticide influences plant concentration from root absorption and translocation. Nutrients influence the penetration of pesticides into plants and translocation of the compounds after they have been absorbed (Talekar and Lichtenstein, 1971).

A detailed account on the plant uptake of pesticides from the soil is beyond the scope of this chapter. The subject matter has been discussed adequately elsewhere (Foy et al., 1971; Nash, 1974; Edwards, 1976).

REFERENCES

Alexander, M. and Aleem, M.I., 1961. J. Agric. Food Chem., 9: 44-47.
Ambrosi, D., Kearney, P.C. and Macchia, J.A., 1977a. J. Agric. Food Chem., 25: 342-347.
Ambrosi, D., Kearney, P.C. and Macchia, J.A., 1977b. J. Agric. Food Chem., 25: 868-872.
Bartha, R., 1971. J. Agric. Food Chem., 19: 385-387.
Bélanger, A. and Hamilton, H.A., 1979. J. Environ. Sci. Health, B14: 213-226.
Beynon, K.I., Davies, L. and Elgar, K., 1967. J. Sci. Food Agric., 17: 167-175.
Booth, G.M., Rhees, R.W., Ferrell, D. and Larsen, J.R., 1975. In: D.D. Kaufman, G.G. Still, G.D. Paulson and S.K. Bandal (Editors), Bound and Conjugated Pesticide Residue. ACS Symposium Ser. 29, pp. 364-365.
Burnside, O.C., Fenster, C.R. and Wicks, G.A., 1971. Weed Sci., 19: 290-293.
Caro, J.H., 1969. Phytopathology, 59: 1192-1197.
Chisaka, H. and Kearney, P.C., 1970. J. Agric. Food Chem., 18: 854-858.
Corbin, F.T. and Upchurch, R.P., 1967. Weeds, 15: 370-376.
Cramp, S. and Olney, P.J.S., 1967. Roy. Soc. Prot. Birds Rept., 1964-1966, 26.
Davis, B.N.K., 1968. Ann. Appl. Biol., 61: 29-45.
Decker, G.C., Bruce, N.W. and Bigger, J.H., 1965. J. Econ. Entomol., 58: 266-271.
Dimond, J.B., Belyea, R.A., Kadunce, R.A., Getchell, A.S. and Blease, J.A., 1970. Can. Entomol., 102: 1122-1130.
Duseja, D.R. and Holmes, E.E., 1978. Soil Sci., 125: 41-48.
Edwards, C.A., 1966. Residue Rev., 13: 83-132.
Edwards, C.A., 1976. Persistent Pesticides in the Environment, CRC Press, Cleveland, Ohio, 170 pp.

Edwards, C.A. and Thompson, A.R., 1973. Residue Rev., 45: 1-79.
Esser, H.O., Dupuis, G., Ebert, E., Marco, G.J. and Vogel, C., 1975 In: P.C. Kearney and D.D. Kaufman (Editors), Herbicides, Vol. 1, Dekker, New York, N.Y., pp. 129-208.
Flashinski, S.J. and Lichtenstein, E.P., 1974. Can. J. Microbiol., 20: 871-875.
Foy, C.L., Coats, G.E. and Jones, D.W., 1971. In: P.L. Altman and D.S. Dittmer (Editors), Respiration and Circulation, Biological Handbooks, Fed. Am. Soc. for Ex. Biol. Bethesda, Md. pp. 743-791.
Fuhremann, T.W. and Lichtenstein, E.P., 1978. J. Agric. Food Chem., 26: 605-610.
Gardiner, J.A., Rhodes, R.C., Adams, J.B. Jr. and Soboczenski, E.J., 1969. J. Agric. Food Chem., 17: 980-986.
Getzin, L.W. and Rosefield, I., 1966. J. Econ. Entomol., 59: 512-516.
Getzin, L.W. and Shanks, C.H. Jr., 1970. J. Econ. Entomol., 63: 52-58.
Gish, C.D., 1970. Pestic. Monit. J., 3: 241-252.
Golab, T. and Althaus, W.A., 1975. Weed Sci., 23: 165-171.
Golab, T. and Amundson, M.E., 1975. Environ. Qual. Safety Suppl. III. 258-261.
Golab, T., Althaus, W.A. and Wooten, H.L., 1979. J. Agric. Food Chem., 27: 163-179.
Golab, T., Herberg, R.J., Gramlich, J.V., Raun, A.P. and Probst, G.W., 1970. J. Agric. Food Chem., 18: 838-844.
Golab, T., Bishop, C.E., Donoho, A.L., Manthey, J.A. and Zornes, L.L., 1975. Pestic. Biochem. Physiol., 5: 196-204.
Goring, C.A.I., 1967. Ann. Rev. Phythol., 5: 285-318.
Graham-Bryce, I.J., 1968. Soc. Chem. Ind. Monograph 29: 251-267.
Hadaway, A.B. and Barlow, F., 1964. Bull. World Health Org., 30: 146-148.
Hamaker, J.W., 1966. Am. Chem. Soc. Adv. Chem. Ser., 60: 122-131.
Hamaker, J.W., 1972. In: C.I. Goring and J.W. Hamaker (Editors), Organic Chemicals in the Soil Environment, Dekker, New York, N.Y., pp. 253-340.
Harris, C.R., Chapman, R.A. and Miles, J.R.W., 1977. J. Environ. Sci. Health, B12: 163-177.
Harris, C.R. and Sans, W.W., 1969. Pestic. Monit. J., 3: 182-185.
Helling, C.S. and Krivonak, A.E., 1978a. J. Agric. Food Chem., 26: 1156-1163.
Helling, C.S. and Krivonak, A.E., 1978b. J. Agric. Food Chem. 26: 1164-1172.
Helling, C.S., Kearney, P.C. and Alexander, M., 1971. Advan. Agron., 23: 147-240.
Hermanson, H.P., Gunther, F.A., Anderson, L.D. and Garber, M.J., 1971. J. Agric. Food Chem., 19: 722-726.
Hill, J.R., 1975. In: D.D. Kaufman, G.G. Still, G.D. Paulson and S.K. Bandal (Editors), Bound and Conjugated Pesticide Residues, ACS Symposium Ser. 29, pp. 358-361.
Hill, G.D., McGahen, J.W., Baker, H.M., Finnerty, D.W. and Bingeman, C.W., 1955. Agron. J., 47: 93-104.
Hiltbold, A.E., 1974. In: W.D. Guenzi (Editor), Pesticides in Soils and Water, Soil Sci. Soc. Am. Inc., publisher, Madison, Wisc., pp. 203-222.
Hsu, T.S. and Bartha, R., 1973. Soil Sci., 116: 444-452.
Hsu, T.S. and Bartha, R., 1976. J. Agric. Food Chem., 24: 119-122.
Isensee, A.R., Holden, E.R., Woolson, E.A. and Jones, G.E., 1976. J. Agric. Food Chem., 24: 1210-1214.

Katan, J. and Lichtenstein, E.P., 1977. J. Agric. Food Chem., 25: 1404-1408.
Katan, J., Fuhremann, T.W. and Lichtenstein, E.P., 1976. Science, 193: 892-894.
Kaufman, D.D., 1967. J. Agric. Food Chem., 15: 582-591.
Kaufman, D.D. 1976. In: D.D. Kaufman, G.G. Still, G.D. Paulson and S.K. Bandal (Editors), Bound and Conjugated Pesticide Residues, ACS Symp. Serv. 29, pp. 1-10.
Kaufman, D.D., Haynes, S.C., Jordan, E.G. and Kayser, A.J., 1977. ACS Symp. Ser. 42, 147-161.
Kazano, H., Kearney, P.C. and Kaufman, D.D., 1972. J. Agric. Food Chem., 20: 975-979.
Kearney, P.C., 1966. Organic Pesticides in the Environment, Adv. Chem. Ser. 60, 250-262.
Kearney, P.C., 1976. In: D.D. Kaufman, G.G. Still, G.D. Paulson and S.K. Bandal (Editors), Bound and Conjugated Pesticide Residues, ACS Symp. Ser. 29, pp. 378-382.
Kearney, P.C., Nash, R.G. and Isensee, A.R., 1969. In: M.W. Miller and G.G. Berg (Editors), Chemical Fallout: Current Research on Persistent Pesticides, Thomas Springfield, Illinois, pp. 54-67.
Khan, S.U. and Marriage, P.B., 1977. J. Agric. Food Chem., 25: 1408-1413.
Khan, S.U. and Marriage, P.B., 1979. Weed Sci., 27: 238-241.
Khan, S.U. and Hamilton, H.A., 1980. J. Agric. Food Chem., (in press).
Khan, S.U., Marriage, P.B. and Saidak, W.J., 1975. Can. J. Soil Sci., 55: 73-75.
Khan, S.U., Marriage, P.B. and Saidak, W.J., 1976. Weed Sci., 24: 583-586.
Khan, S.U., Hamilton, H.A. and Hogue, E.J., 1976. Pestic. Sci., 7: 553-558.
Khan, S.U., Bélanger, A., Hogue, E.J., Hamilton, H.A. and Mathur, S.P., 1976. Can. J. Soil Sci., 56: 407-412.
Lichtenstein, E.P., 1966. Nat. Acad. Sci. Nat. Res. Counc. Publ., 1402: 221-229.
Lichtenstein, E.P., 1966. J. Econ. Entomol., 59: 985-993.
Lichtenstein, E.P., 1972. In: Pesticide Chemistry, Proceedings, 2nd International IUPAC Congress, Vol. VI, A.S. Tahon (Editor) Gordon and Breach, London.
Lichtenstein, E.P. and Schulz, K.R., 1964. J. Econ. Entomol., 57: 618-627.
Lichtenstein, E.P., Myrdal, G.R. and Schulz, K.R., 1965a. J. Agric. Food Chem., 13: 126-131.
Lichtenstein, E.P., Schulz, K.R., Skrentny, R.F. and Shitt, P.A., 1965b. J. Econ. Entomol. 58: 742-746.
Lichtenstein, E.P., Katan, J. and Anderegg, B.N., 1977. J. Agric. Food Chem., 25: 43-47.
Marriage, P.B., Khan, S.U. and Saidak, W.J., 1977. Weed Res., 17: 219-225.
Marriage, P.B., Saidak, W.J. and Von Stryk, F.G., 1975. Weed Res., 15: 373-379.
Menzer, R.E., Fontanilla, E.L. and Ditman, L.P., 1970. Bull. Environ. Contam. Toxicol. 5: 1-5.
Menzie, C.M., 1972. Ann. Rev. Entomol., 17: 199-222.
Miles, J.R.W., 1968. J. Agric. Food Chem., 16: 620-622.
Miles, J.R.W. and Harris, C.R., 1978a. J. Econ. Entomol., 71: 125-131.
Miles, J.R.W. and Harris, C.R., 1978b. J. Environ. Sci. Health, B13: 199-209.
Miles, J.R.W., Harris, C.R. and Moy, P., 1978. J. Econ. Entomol., 71: 97-101.

Muir, D.C.G. and Baker, B.E., 1978. Weed Res., 18: 111-120.
Nash, R.G., 1974. In: W.D. Guenzi (Editor), Pesticides in Soil and Water, Soil Sci. Soc. Amer. Inc., Madison, Wisc., pp. 257-313.
Nash, R.G. and Woolson, E.A., 1967. Science, 157: 924-927.
Nash, R.G., Harris, W.G. and Ensor, P.D., 1973. J. Assoc. Official Anal. Chem., 56: 728-732.
Oloffs, P.C., Szeto, S.Y. and Webster, J.M., 1971. Can. J. Plant Sci., 51: 547-550.
O'Toole, M.A., 1966. Weed Abstr., 15: 58.
Otto, S. and Drescher, N., 1973. Lab Report 1143, Badische Anilin and Soda Fabrik AG (BASF).
Owen, R.B. Jr., Dimond, J.B. and Getchell, A.S., 1977. J. Environ. Qual., 6: 359-360.
Parks, S.J. and Tepe, J.B., 1969. Weed Sci., 17: 119-122.
Phillips, W.M., 1968. Weed Sci., 16: 144-148.
Probst, G.W., Golab. T., Herberg, R.J., Holzer, F.J., Parks, S.J., Van der Schans, C., Tepe, J.B., 1967. J. Agric. Food Chem., 15: 592-599.
Rhodes, R.C., 1977. J. Agric. Food Chem., 25: 529-533.
Saha, J.G. and Sumner, A.K., 1971. Pestic. Monit. J., 5: 28-31.
Saha, J.G., Craig, C.H. and Janzen, W.K., 1968. J. Agric. Food Chem., 16: 617-619.
Saha, J.G., Burrage, R.H., Lee, Y.W., Saha, M. and Sumner, A.K., 1974. Can. J. Plant Sci., 54: 717-723.
Savage, K.E., 1978. Weed Sci., 26: 465-471.
Sheets, T.J., 1967. In: Agriculture and the Quality of Our Environment, A.A.A.S. Publ., 85: 20.
Sheets, T.J., 1970. Residue Rev., 32: 287-310.
Sheets, T.J. and Harris, C.R., 1965. Residue Rev., 11: 119-140.
Sheets, T.J., Crafts, A.S. and Drever, H.R., 1962. J. Agric. Food Chem., 10: 458-462.
Sheets, T.J., Smith, J.W. and Kaufman, D.D., 1968. Weed Sci., 16: 217-222.
Sirons, G., Frank, R. and Sawyer, T., 1973. J. Agric. Food Chem., 21: 1016-1020.
Smith, R.A., Belles, W.S., Shen, K.W. and Woods, W.G. 1973. Pestic. Biochem. Physiol., 3: 278.
Spencer, E.Y., 1968. Canada Dept. Agr. Publ., 1093. 5th ed. p. 483.
Spillner, C.J. Jr., DeBaun, J.R. and Menn, J.J., 1979. J. Agric. Food Chem., 27: 1054-1060.
Steevens, D.R., Walsh, L.M. and Keeney, D.R., 1972. Pest. Mon. J., 6: 89-90.
Stevens, L.J., Collier, C.W. and Woodham, D.W., 1970. Pest. Mon. J., 4: 145.
Suss, A. and Grampp, B., 1973. Weed Res., 13: 254-266.
Swan, D.G., 1972. Weed Sci., 20: 335-337.
Tafuri, F., Businelli, M., Scarponi, L. and Marucchimi, C., 1977. J. Agric. Food Chem., 25: 353-356.
Talekar, N.S. and Lichtenstein, E.P., 1971. J. Agric. Food Chem., 19: 846-850.
Thompson, A.R., 1973. In: C.A. Edwards (Editor), Environmental Pollution by Pesticides, Plenum Press, New York, N.Y., pp. 87-133.
Tsukamoto, M. and Suzuki, R., 1964. Botyu - Kagaku, 29: 76-89.
Upchurch, R.P., 1966. Residue Rev., 16: 46-85.
Upchurch, R.P., Corbin, F.T. and Selman, F.L., 1969. Weed Sci., 17: 69-77.
U.S. Environmental Protection Agency, 1975. Fed. Regist., 40(123), 26802.

Van Alfen, N.K. and Kosuge, T.J., 1976. J. Agric. Food Chem., 24: 584-588.
Viswanathan, R., Scheunert, I., Kohli, J., Klein, W. and Korte, F., 1978. J. Environ. Sci. Health, B13: 243-259.
Walker, A. and Smith, A.E., 1979. Pestic. Sci., 10: 151-157.
Walter-Echols, G. and Lichtenstein, E.P., 1978. J. Agric. Food Chem., 26: 599-604.
Waters, W.E. and Burgis, D.S., 1968. Weed Sci., 16: 149-151.
Wheatley, G.A. and Hardman, J.A., 1965. Nature (London), 207: 486.
Wheatley, G.A. and Hardman, J.A., 1968. J. Sci. Food Agric. 19: 219-225.
Wheeler, W.B., Stratton, G.D., Twilley, R.R., Ou, Li-Tse, Carlson, D.A. and Davidson, J.M., 1979. J. Agric. Food Chem., 27: 702-706.
Wiese, A.F., Chenault, E.W. and Hudspeth, E.B. Jr., 1969. Weed Sci., 17: 481-483.
Williams, J.H., 1975. Pestic. Sci. 6: 501-509.
Williams, I.H. and Brown, M.J., 1979. J. Agric. Food Chem., 27: 130-132.

Chapter 6

MINIMIZING PESTICIDE RESIDUES IN SOIL

The surest way to avoid pesticide residues in soils is to stop using these chemicals for crop protection and pest control. This choice is not open from a practical standpoint. For an efficient food production to support the rapidly expanding world population, it appears likely that the use of pesticides will continue to increase in the foreseeable future. Our aim, therefore, must be to minimize the undesirable environmental consequence of the use of pesticides.

6.1. ALTERNATIVE TO PESTICIDES

Several alternative approaches to crop protection and pest control have been used with varying degrees of success. Prior to the development of modern pesticides, man had widely used cultivation practices and plant breeding as traditional methods. Weeds have been controlled by a careful preparation of the seed bed by ploughing, mechanical weeding and hand hoeing. Methods such as timing of sowing dates, timing of harvesting, crop rotation and the use of crops resistant to disease and pests have long been known.

Integrated control is a relatively new concept for pest management (Apple and Smith, 1976). Smith (1977) offered the following description of integrated pest control: Integrated pest control is a multidisciplinary, ecological approach to the management of a pest population, which utilizes a variety of control tactics compatibly in a single coordinated pest management system. In its operation, integrated pest control is a multi-tactical approach that encourages the fullest use of natural mortality factors complemented when necessary by artificial means of pest management. Also implicit in its definition is the understanding that imposed artificial control measures, notably convention pesticides, should be used only where economic injury thresholds would otherwise be exceeded. As a corollary to this, integrated pest control is not dependent on any single control procedure or tactic. For each situation, the strategy is to coordinate the relevant tactics with the natural regulating and limiting elements of the environment. Thus, it implies that when control measures are used, they should integrate cultural and ecological control measures with pesticidal ones to obtain the maximum effect with a minimum use of pesticides. Application of this principle may reduce the amount of pesticides being used without decreasing levels of effectiveness or increasing loss.

Biological control of weeds and pests has attracted attention for many years (Huffaker, 1977; DeBach, 1970). Introduction of insects that feed specifically on certain weeds has been found useful. The cochineal insect, *Dactylopius tomentossus* and the moth *Cactoblastis caetorum* control the prickly pear. This technique can therefore be used for the long term control of a single dominant weed present over large areas of uncropped land. However, the method will have serious limitations for a rapid control of mixed weed infestations. The introduced insect may also attack related plants of economic importance and produce adverse effects on the natural ecological balance in the area.

Control of insects by predators, parasites or pathogens can be a cheap method of crop protection. One of the more promising newer methods of utilizing parasites and/or predators is the inundative release of a beneficial insect to reduce the population of the pest before it reaches a damaging level. Pathogens for control of noxious insects have received increased attention. Biological control methods have usually been successful only with imported predators or parasites to control imported pests and in areas isolated topographically or geographically. If this technique is to be used, a very large number of predators or parasites are required. Production of these large numbers may present considerable difficulties. Furthermore, even if a pest predator is established successfully, it becomes essential not to use any pesticide that will kill the predator. Wilson (1970) pointed out that altogether there have been more than 220 examples of successful biological control involving 110 species of pests in more than 60 countries.

The sterile male technique has been found very useful in eradicating the screw worm from the south eastern United States. A great deal of interest has also been expressed in the use of various genetic methods of insect control. Within the past few years a great deal of work has been done on insect pheromones. Insect hormones, which regulate development, are also being tested for the control of a number of noxious insects.

Biological control agents have the advantage of being highly specific in that they affect only the target pest. However, there is a possibility that the introduced insect might become a pest of some economic crop. An introduced pathogen might change and become infectious to man or animals. Thus, the possibility exists that biological pest control may become an environmental risk. It appears that while all the nonchemical methods will play their part alongside pesticides, the latter will be the maintstay for many years to come until equally alternative methods are found for plant protection and pest control.

6.2. SHORT RESIDUAL PESTICIDES

The main problem of finding short residual pesticides as an alternative for persistent pesticides is not that a suitable compound cannot be developed. It is because the persistent pesticides can be made and sold very cheaply, and often eliminate the need for repeated applications. The possibilities of producing biodegradable analogues of organochlorine insecticides were investigated by Metcalf et al. (1972). They followed the breakdown of one possible biodegradable analogue, methoxychlor, and showed that all its metabolites degraded readily. It is possible to replace most persistent organochlorines with biodegradable analogues (Metcalf, 1971). Possible biodegradable analogues of DDT include methoxychlor, ethoxychlor, methylchlor, and methiochlor.

6.3. ELIMINATING PESTICIDE RESIDUES

Complete elimination of some of the persistent pesticide residues from soils may be impractical or even impossible. However, a lowering of existing residues may result in minimal residues in crops grown on the soil. This can be achieved by planting a tolerant crop. Occasionally crop cultivars can be bred for tolerance to a specific pesticide (Williams and Johnson, 1953). Plants which have an affinity for pesticides could be grown on contaminated soil and then removed after their having taken up some part of the presidual pesticide. Deep plowing could be used to incorporate a pesticide into the soil and in eradicating it from the surface soil. This practice may not be desirable as the pesticide degradation may be reduced considerably in the subsoil (Roeth et al. 1969; Harris et al., 1969). However, it has been observed that residues of nitralin and trifluralin in the soil surface layer can be made nontoxic to a susceptible crop by plowing the soil before planting (Burnside, 1972). Irrigation can leach a pesticide out of the root zone so that crops could be grown on the land (Lange, 1970). Yoshida and Castro (1970) observed that in a flooded sandy loam soil no lindane remained after one month. The use of an adsorbant, such as activated carbon has received considerable attention in the detoxification of pesticides (Foy and Bingham, 1969). Chemical and microbial additions have also been shown to detoxify certain pesticides. The disappearance of DDT occurred most rapidly in soils inoculated with *Aerobacter erogenes*, under flooded, anaerobic soil conditions (Kearney et al. 1969). Surfactants have been found useful in regulating the depth of penetration and persistence of pesticides (Bayer, 1967).

Biological and nonbiological means for dissipating pesticide residues in soils have also been investigated (Alexander, 1967; Sheets and Kaufman, 1970). Nonbiological means include adsorption, leaching, volatilization, photodecomposition and various other chemical reactions. The biological means of pesticide dissipation are mitigated by higher plants and microorganisms.

6.4. FUTURE NEEDS

The use of pesticides for crop protection and pest control will be continued and intensified until equally effective alternative methods are found. Pesticides that present unacceptable degrees of residue risk should be replaced by alternative pesticides, which will cause less residual and environmental hazard and will fit in best with established agricultural practices. Consequently, there will be a continuing need for research and development of new pesticides.

For a maximum effective use of agricultural land, herbicides will have a large part to play in providing weed control needed to increase crop yield. Discovery and development of new herbicides will mainly be of those of the conventional type. There is a need to undertake additional studies of specific fundamental reactions of herbicides in soil and to establish the degree to which these reactions are of consequences under various use conditions in the field.

Recently much attention has been given to improving conventional insecticides. The organochlorines are not easily broken down in soil and have therefore persisted for a long time. On the other hand organophosphates and carbamates break down rapidly in soil and are therefore not persistant and present no long term residual effect. We should attempt to find out how the more persistent insecticides behave and break down in soil, so that their persistance and pathway of breakdown can be predicted in the future. The potential of the new synthetic pyrethroids has yet to be assessed. They appear not to be persistant in the environment.

The standard protective fungicides such as sulfur, copper preparations and dithiocarbamates are likely to continue to be widely used. Future research could be directed to further develop new fungicides effective against soil borne diseases and which can be used by their incorporation into the soil at the time of sowing. There is also a need for a wide range of systemic fungicides whose effectiveness can be maintained by restrained use and rotation.

REFERENCES

Alexander, M., 1967. In: Agriculture and the Quality of Our Environment, N.C. Brady (Editor), Amer. Assoc. Advan. Sci. Publ. 85, Washington, D.C., pp. 331-342.

Apple, J.L. and Smith, R.F. (Editors), 1976. Integrated Pest Management, Plenum, New York, N.Y., 200 pp.

Bayer, D.E., 1967. Weeds, 15: 249-252.

Burnside, O.C., 1972. Weed Sci., 20: 294-297.

DeBach, P., (Editor), 1970. Biological Control of Insect Pests and Weeds, Chapman and Hall, London, 844 pp.

Foy, C.L. and Bingham, S.W., 1969. Residue Rev., 29: 105-135.

Harris, C.I., Woolson, E.A. and Hummer, B.E., 1969. Weed Sci., 17: 27-31.

Huffaker, C.B. (Editor), 1977. Biological Control, Plenum, New York, N.Y., 511 pp.

Kearney, P.C., Woolson, E.A., Plimmer, J.R. and Isensee, A.R., 1969. Residue Rev., 29: 137-149.

Lange, A.H., 1970. Proc. West Weed Contr. Conf. 27: 30.
Metcalf, R.L., 1971. J. Soil Water Conserv., 26: 57-60.
Metcalf, R.L., Kapoor, I.P. and Hirwe, A.S., 1972. Chemtech, February, 105-109.
Roeth, F.W., Lavy, T.L. and Burnside, O.C., 1969. Weed Sci., 17: 202-205.
Sheets, T.J. and Kaufman, D.D., 1970. In: FAO International Conference on Weed Control, Weed Sci. Soc. Am., Urbana, Illinois, pp. 513-538.
Smith, R.F., 1977. In: E.H. Smith and D. Pimental (Editors), Pest Control Strategies, National Inform. Servic. Rep., PB-274 644, Springfield, Va., pp. 41-81.
Williams, J.H. and Johnson, I.J., 1953. Agron. J., 45: 298-301.
Wilson, F., 1970. Adv. Sci., 26: 374-378.
Yoshida, T. and Castro, T.F., 1970. Soil Sci. Soc. Am. Proc., 34: 440-442.

APPENDIX

TABLE A-1

Listing of pesticides referred to in text by common names, other names and chemical names[1]

Common name	Other name	Class	Chemical name
Agvitor		I	2,4,5-trichlorophenyl diethyl-phosphinothionate
Alachlor	Lasso	H	α-chloro-2,6-diethyl-*N*-methoxy-methylacetanilide
Aldicarb	Temik	I,N	2-methyl-2-(methylthio)propion-aldehyde *O*-(methylcarbamoyl)oxime
Aldrin	HHDN	I	1,2,3,4,10,10-hexachloro-1,4a,4,5,8,8a-hexahydro-*exo*-1,4-*endo*-5,8,-dimethanonaphthalene
Ametryn	Evik, Gesapax	H	2-methylthio-4-(ethylamino)-6-(isopropylamino)-*s*-triazine
Amiprophos		H	ethyl-2-nitro-4-methyl *N*-isopropylphosphoramidothionate
Amitrole	Aminotri-azole, Amizol	H,M	3-Amino-*s*-triazol
Arsenic trioxide		H	
Atrazine	AAtrex	H	2-chloro-4-(ethylamino)-6-(iso-propylamino)-*s*-triazine
Azinphos-methyl	Guthion	I,A	*S*-(3,4-dihydro-4-oxobenzo[d]-[1,2,3]triazin-3-ylmethyl) *O*,*O*-dimethyl phosphorodithioate
Barban	Carbyne	H	4-chlorobut-2-ynyl 3-chlorophenyl-carbamate
Benfluralin	Benefin, Balan	H	*N*-butyl-*N*-ethyl-2,6-dinitro-4-trifluoromethyl aniline
Benomyl	Benlate, Tersan	F	methyl 1-(butylcarbamoyl) benzimidazol-2-ylcarbamate
Bensulide	Betasan	H	*O*,*O*-di-isopropyl *S*-2-phenylsulpho-nylaminoethyl phosphorodithioate
Bentazon	Basagran	H	3-isopropyl-(1*H*)-benzo-2,1,3-thia-diazin-4-one 2,2-dioxide

TABLE A-1 (*continued*)

Common name	Other name	Class	Chemical name
Bromacil	Hyvarx	H,A	5-bromo-3-*sec*-butyl-6-methyluracil
Bromophos	Nexion	I	*O*-(4-bromo-2,5-dichlorophenyl) *O,O*-dimethyl phosphorothioate
Bromoxynil	Brominil, Buetril	H	3,5-dibromo-4-hydroxy-benzonitril
Butralin	Amex 820, Dibutalin	H	*N*-*sec*-butyl-4-*tert*-butyl-2,6-dinitroaniline
Cacodylic acid	Phytar 138, Chexmate	H	Hydroxydimethylarsine oxide
Captan	SR 406, Orthocide 406	F	3a,4,7,7a-tetrahydro-*N*-(trichloromethanesulphenyl)-phthalimide
Carbaryl	Sevin	I,P	1-naphthyl methylcarbamate
Carbendazim	MBC,MCAB, BCM	F	methyl benzimidazol-2-ylcarbamate
Carbofuran	Furadan	I,A,N	2,3-dihydro-2,2-dimethyl benzofuran-7-yl methylcarbamate
Carbon disulphide	carbon bisulphide	Fu	
Carbophenothion	Trithion	I,A	*S*-4-chlorophenylthiomethyl *O,O*-diethyl phosphorodithioate
Carboxin	Vitavax	Fu	5,6-dihydro-2-methyl-1,4-oxatiin-3-carboxanilide
CDAA	Randox	H	*N,N*-diallyl-2-chloroacetamide
CDEC	Vegadex	H	2-chloroallyl diethyldithiocarbamate
Chloramben	Amiben	H	3-amino-2,5-dichlorobenzoic acid
Chloranil	Spergon	F	2,3,5,6-tetrachloro-*p*-benzoquinone
Chlordane	Chlordan	I	1,2,4,5,6,7,8,8-octachloro-3a,4,7,7a-tetrahydro-4,7-methanoindane
Chlorfenvinphos	Birlane, Supona	I	2-chloro-1-(2,4-dichlorophenyl) vinyl diethyl phosphate
Chloroneb	Tersan, Demosan	F	1,4-dichloro-2,5-dimethoxybenzene

TABLE A-1 (*continued*)

Common name	Other name	Class	Chemical name
Chlorphen-amidine	Chlord-imeform	I,A	*N*-(4-chloro-*O*-tolyl)-*N*,*N*-dimethyl-formamidine
Chloropicrin		I,Fu,N	trichloronitromethane
Chloroxuron	Tenoran	H	3-[4-(4-chlorophenoxy)phenyl]-1,1-dimethylurea
Chlorpro-pham	Furloe	H	isoproyl *m*-chlorocarbanilate
Chlorpy-rifos	Dursban®	I	*O*,*O*-diethyl *O*-3,5,6-trichloro-2-pyridyl phosphorothiate
Chlort-hiamid	Prefix	H	2,6-dichlorothiobenzamide
Chlorto-luron	Dicuron	H	3-(3-chloro-*p*-tolyl)-1,1-dimethylurea
Crotoxy-phos	Ciodrin	I	dimethyl *Z*-1-methyl-2-(1-phenylethoxycarbonyl)vinyl phosphate
Cypermet-hrin	*cis*-isomer NRDC 160 *trans*-isomer NRDC 159	I	α-cyano-3-phenoxybenzyl(±)*Z*, *E*-3-(2,2-dichlorovinyl)-2,2-dimethylcyclopropane carboxylate
2,4-D		H	2,4-dichlorophenoxyacetic acid
Dalapon	Dowpon	H	2,2-dichloropropionic acid
2,4DB	Embutox	H	4-(2,4-dichlorophenoxy)butyric acid
DCPA	Dacthal	H	dimethyl tetrachloroterephthalate
2,4-DEP	Falon	H	tris[2-(2,4-dichlorophenoxy) ethyl phosphite
DBH			dichlorobenzhydrol
DBP			dichlorobenzophenone
DDA			dichlorodiphenylacetic acid
DDCN			dichlorodiphenylacetonitrile
DDD			dichlorodiphenyldichloroethane
DDE			dichlorodiphenyldichloroethylene

TABLE A-1 (*continued*)

Common name	Other name	Class	Chemical name
DDM			dichlorodiphenylmethane
DDMU			dichlorodiphenylchloroethylene
DDMS			dichlorodiphenylchloroethane
DDNS			dichlorodiphenylethane
DDNU			dichlorodiphenylethylene
DDOH			dichlorodiphenylethanol
DDT		I	a technical mixture of isomers of 1,1,1-trichloro-2,2-bis(*p*-chlorophenyl)ethane,*p*,*p*-DDT predominates (>70% w/w)
p,*p*-DDT		I	1,1,1-trichloro-2,2-bis(*p*-chlorophenyl)ethane
Dicofol			dichlorodiphenyltrichloroethanol
Demeton-*O*	Systox	I,A	*O*,*O*-diethyl *O*-(2-ethylthioethyl) phosphorothioate
Dexon	Fenaminsulf	F	sodium *p*-(dimethylamino)benzenediazo
Diallate	Avadex	H	*S*-(2,3-dichloroallyl)diisopropylthiocarbamate
Diazinon	Basudin	I	*O*,*O*-diethyl *O*-2-isopropyl-6-methyl pyrimidin-4-yl phosphorothioate
Dibromochloropropane	DBCP, Fumazone, Nemagon	Fu	1,2-dibromo-3-chloropropane
Dicamba	Banvel D	H	3,6-dichloro-2-methoxybenzoic acid
Dichlobenil	Casoran	H	2,6-dichlorobenzonitrile
Dichlofenthion	VC-13	I,N	*O*-(2,4-dichlorophenyl)*O*,*O*-diethyl phosphorothioate
Dichlormate	Rowmate	H	3,4-dichlorobenzyl methylcarbamate
Dichloropropene mixture	D-D	Fu,N	mixture of (*E*)-and (*Z*)-1,3-dichloropropene

TABLE A-1 *(continued)*

Common name	Other name	Class	Chemical name
Dichlorvos	DDVP, Vapona	I,A	2,2-dichlorovinyl dimethyl phosphate
Dicrotophos	Bidrin	I	dimethyl *Z*-2-dimethylcarbamoyl-1-methyl vinyl phosphate
Dicryl	Chloranocryl	H	*N*-(3,4-dichlorophenyl) methacrylamide
Dieldrin		I	1,2,3,4,10,10-hexachloro-6,7-epoxy-1,4,4a,5,6,7,8,8a-octahydro-*exo*-1,4-*endo*-5,8-dimethanonaphthalene
Dimefox	Terrasytam	I	bis(dimethylamino)fluorophosphine oxide
Dimethoate	Cygon	I,A	*O,O*-dimethyl *S*-methylcarbamoylmethyl phosphorodithioate
Dinitramine	Cobex	H	*N,N*-diethyl-2,6-dinitro-4-trifluoromethyl-*m*-phenylenediamine
Dinosam	DNAP	I,A,N	2-(1-methylbutyl)-4,6-dinitrophenol
Dinoseb	DNBP	I,A,H	2-*sec*-butyl-4,6-dinitrophenol
Diphenamid	Dymid	H	*N,N*-dimethyl-diphenylacetamide
Diquat	Reglone	H	1,1-ethylene-2,2-dipyridylium di-ion
Disodium methanearsenic acid	DSMA	H	
Disulfoton	Di-Syston	I,A	*O,O*-diethyl *S*-(2-ethyl-thio-ethyl) phosphorodithiate
Diuron	DMU	H	3-(3,4-dichlorophenyl)-1,1-dimethylurea
DMPA	Zytron	H	2,4-dichlorophenyl methyl *N*-isopropylphosphoramidothionate
DNOC	DNC	I,A	4,6-dinitro-*O*-cresol
DSMA	Ansar-8100	H	disodium methanearsonate
Endosulfan	Thiodan	I,A	6,7,8,9,10-hexachloro-1,5,5a,6,9,9a-hexahydro-6,9-methano-2,4,3-benzo(*e*)dioxathiepin 3-oxide

TABLE A-1 (*continued*)

Common name	Other name	Class	Chemical name
Endothall	Endothal-sodium	H	7-oxabicyclo[2,2,1]heptane-2,3-dicarboxylic acid
Endrin		I	1,2,3,4,10,10-hexachloro-6,7-epoxy-1,4,4a,5,6,7,8,8a-octahydro-*exo*-1,4-*exo*-5,8-dimethanonaphthalene
EPBP	S-Seven	I	2,4-dichlorophenyl ethyl phenyl-phosphonothionate
EPTC	Eptam	H	*S*-ethyl dipropylthiocarbamate
Ethion	Nialate	A,I	*O,O,O,O*-tetraethyl *S-S*-methylene di(phosphorodithioate)
Ethirimol	Milstem	F	5-butyl-2-ethylamino-4-hydroxy-6-methyl pyrimidine
Ethylene Dibromide	EDB	Fu	1,2-dibromoethane
ETU		Fu	ethylene thiourea
Fenac		H	2,3,6-Trichlorophenyl acetic acid
Fenitrothion	Sumithion	I	*O,O*-dimethyl *O*-3-methyl-4-nitrophenyl phosphorothioate
Fensulfothion	Dasanit, Terracur P	I,N	*O,O*-diethyl *O*-(4-methylsufinyl-phenyl)phosphorothioate
Fenthion	Baytex	I,A	*O,O*-dimethyl *O*-4-methylthio-*m*-tolyl)phosphorothiate
Fenuron	Dybar	H	1,1-dimethyl-3-phenylurea
Fenvalerate	WL 43775, Pydrin	I	α-cyano-3-phenoxybenzyl 2-(4-chlorophenyl)-3-methylbutyrate
Fluchloralin	Basalin	H	*N*-(2-chloroethyl)-2,6-dinitro-*N*-propyl-4-trifluoromethylaniline
Fluometuron	Cotoran	H	1,1-dimethyl-3(3-trifluoro-methylphenyl)urea
Fonofos	Dyfonate	I	*O*-ethyl *S*-phenyl ethylphosphono-dithioate
Formaldehyde	Formalin	Fu	

TABLE A-1 (*continued*)

Common name	Other name	Class	Chemical name
Glyphosate	Roundup	H	*N*-(phosphonomethyl)glycine
Heptachlor		I	1,4,5,6,7,8,8-heptachloro-3a,4,7,7a-tetrahydro-4,7-methanoindene
Hexachloro-orobenzene		I	
Isodrin		I	hexachlorohexahydro-*endo,endo*-dimethanonaphthalene
Ioxynil	Totril	H	4-hydroxy-3,5-diiodobenzonitrile
Ipazine	Gesabal	H	2-chloro-4-diethylamino-6-isopropylamino-*s*-triazine
Isobenzan	Telodrin	I	1,3,4,5,6,7,8,8-octachloro-1,3,3a,4,7,7a-hexahydro-4,7-methanoisobenzofuran
Isocil	Hyvar	H	5-bromo-3-isopropyl-6-methyluracil
Leptophos	Phosvel	I	*O*-(4-bromo-2,5-dichlorophenyl) *O*-methyl phenylphosphonothioate
Lindane	gamma-BHC, gamma-HCH	I	1,2,3,4,5,6-hexachlorocyclohexane
Linuron	Lorox	H	3-(3,4-dichlorophenyl)-1-methoxy--1-methylurea
Malathion	Cythion	I,A	*S*-1,2-di(ethoxycarbonyl)ethyl *O*-*O*-dimethyl phosphorodithioate
Maneb	Manzate	F	manganese ethylenebisdithiocarbamate
MCPA	Agroxone	H	4-chloro-2-methylphenoxyacetic acid
Mecarbam	Murfotox	I,A	*S*-(*N*-ethoxycarbonyl-*N*-methylcarbamoyl methyl)*O*,*O*-diethyl phosphorodithioate
Metacrephos	Cremart, S-2846	I	ethyl 3-methyl-6-nitrophenyl *N*-*sec*-butylphosphoramidothionate
Metobromuron	Patoran	H	3-(4-bromophenyl)-1-methoxy-1-1-methylurea
Methabenzthiazuron	Tribunil	H	1-benzothiazole-2-yl-1,3-dimethylurea

TABLE A-1 (*continued*)

Common name	Other name	Class	Chemical
Metham	Vapam	F,N,H	sodium methyldithiocarbamate
Methida-thion	Supracide	I	*S*-(2,3-dihydro-5-methoxy-2-oxo-1,3,4-thiadoxol-3-ylmethyl) *O*,*O*-dimethyl phosphorodithioate
Methio-carb	Mesurol	I,A	4-methylthio-3,5-xylylmethyl-carbamate
Methomyl	Lannate	I	*S*-methyl-*N*-(methylcarbamoyloxy) thioacetimidate
Methyl Bromide	Bromoethane	I	
Methoxy-chlor	Methoxy-DDT	I	1,1,1-trichloro-2,2-di-(4-methoxyphenyl)ethane
Methyl-mercury Dicyandiam-mide	Panogen	F	3-(methylmercurio)guanidino-carbonitrile
Mevinphos	Phosdrin	I,A	2-methoxycarbonyl-1-methyl vinyl dimethyl phosphate
Mocap	Ethoprophos	I,N	*O*-ethyl *S*,*S*-dipropyl phosphoro-dithioate
Molinate	Ordram	H	*S*-ethyl *N*,*N*-hexamethylene-thiocarbamate
Monolin-uron	Aresin	H	3-(4-chlorophenyl)-1-methoxy-1-methylurea
Monuron	Telvar	H	3-(4-chlorophenyl)-1,1-dimethyl-urea
Morestan	Chino-methionat	I,F,A	6-methyl-quinoxaline-2,3-dithiolcyclocarbonate
MSMA	Ansar 170, Trans-Vert	H	monosodium methanearsonate
Naptalam	Alanap	H	*N*-1-naphthylphthalamic acid
Neburon	Kloben	H	1-butyl-3-(3,4-dichlorophenyl)-1-methylurea
Nitralin	Planavin	H	4-methylsulphonyl-2,6-dinitro-*N*,*N*-dipropylaniline
Norea	Herban	H	3-(hexahydro-4,7-methanoindan--5-yl)-1,1-dimethylurea

TABLE A-1 (*continued*)

Common name	Other name	Class	Chemical name
Oryzalin	Ryzelan	H	3,5-dinitro-N^4,N^4-dipropylsulfanilamide
Oxadiazon	Ronstar	H	5-*tert*-butyl-3-(2,4-dichloro-5-isopropoxyphenyl)-1,3,4-oxadiazol-2-one
Oxamyl	Thioxamyl	I,N	*N*,*N*-dimethyl-α-methylcarbamoyloximino-α-(methylthio)acetamide
Oxycarboxin	Plantvax	F	2,3-dihydro-6-methyl-5-phenylcarbamoyl-1,4-oxathiin-4,4-dioxide
Oxydisulfoton	Disyston-S	I,A	*O*,*O*-diethyl *S*-(2-ethylsulphinylethyl)phosphorodithioate
Panogen®	MEMA	F	Methoxyethylmercury acetate
Paraquat	Gramoxone, Weedol	H	1,1-dimethyl-4,4-dipyridylium di-ion
Parathion	Folidol	I,A	*O*,*O*-diethyl *O*-(4-nitrophenyl) phosphorothioate
Parathion methyl	Folidol-M	I,A	*O*,*O*-dimethyl *O*-(4-nitrophenyl) phosphorothioate
PCBA			*p*-chlorobenzoic acid
PCP	Penta	H	pentachlorophenol
PCPA			*p*-chlorophenylacetic acid
PCNB		F	1,2,3,4,5-pentachloronitrobenzene
Pebulate	Tillam	H	*S*-propyl butylethylthiocarbamate
Permethrin	NRDC-143, Ambush	I	3-phenoxybenzyl (±) *Z*, *E*-3-(2,2-dichlorovinyl)-2,2-dimethylcyclopropanecarboxylate
Phenthoate	Cidial	I,A	*S*,α-ethoxycarbonylbenzyl *O*,*O*-dimethyl phosphorodithioate
Phenylmercury Acetate	PMA	F	(acetato-*O*)phenylmercury
Phorate	Thimet	I,A	*O*,*O*-diethyl *S*-(ethylthio)methyl phosphorodithioate

TABLE A-1 *(continued)*

Common name	Other name	Class	Chemical name
Phosalone	Zolone	I,A	*S*-6-chloro-2-oxobenzoxazolin-3-yl methyl *O,O*-diethyl phosphorodithioate
Phosfolan	Cyolane	I	diethyl 1,3-dithiolan-2-ylidenephosphoramidate
Picloram	Tordon	H	4-amino-3,5,6-trichloro-picolinic acid
Pirimicarb	Pirimor, Aphox	I	5,6-dimethyl-2-dimethyl-amino-4-pyrimidinyl-dimethylcarbamate
Profenofos	Selecron	I	*O*-(4-bromo-2-chlorophenyl)*O*-ethyl *S*-propyl phosphorothiate
Profluralin	Pregard, Tolben	H	*N*-cyclopropylmethyl-2,6-dinitro-*N*-propyl-4-trifluoromethyl-aniline
Prometone	Primatol, Carbamult	H	2,4-di(isopropylamino)-6-methoxy-*s*-triazine
Prometryn	Gesagard, Caparol	H	2-methylthio-4,6-bis(isopropyl-amino)-*s*-triazine
Pronamide	Kerb	H	*N*(1,1-dimethylpropynyl)3,5-dichlorobenzamide
Propachlor	Ramrod	H	α-chloro-*N*-isopropylacetanilide
Propanil	Rogue, Stam F	H	3,4-dichloropropionanilide
Propazine	Primatol, Milogard	H	2-chloro-4,6-di(isopropyl-amino)-*s*-triazine
Propham	Chem-hoe	H	isopropyl phenylcarbamate
Pyrazon	Pyramin, Alicap	H	5-amino-4-chloro-2-phenyl-pyridazin-3-one
Pyrichlor	Dextron	H	2,3,5-trichloro-4-pyridinol
Quintozene	Braasicol, PCNB	F	pentachloronitrobenzene
Ronnel	Fenchlo-rphos	I,A	*O,O*-dimethyl *O*-(2,4,5-trichloro-phenyl)phosphorothioate
Semesan		Fu	Hydroxymercurichlorophenol
S-5439		I	3-phenoxybenzyl-3-methyl-2-(4-chloro)phenyl butyrate
Semesan		Fu	Hydroxymercurichlorophenol

TABLE A-1 (*continued*)

Common name	Other name	Class	Chemical name
Sesone	Sesone	H	2-(2,4-dichlorophenoxy)ethyl sodium sulfate
Siduron	Tupersan	H	1-(2-methylcyclohexyl)-3-phenylurea
Simazine	Gesatop, Primatol	H	2-chloro-4,6-di(ethylamino) -*s*-triazine
Simetone		H	2-methoxy-4,6-di(ethylamino) -*s*-triazine
Sodium Arsenite	ARCADIAN Sodium Arsenite "8" Solution	H	
Solan	Pentanochlor	H	3-chloro-4-methyl-α-methylvaler-anilide
Swep		H	methyl 3,4-dichlorocarbanilate
2,4,5-T		H	2,4,5-trichlorophenoxyacetic acid
2,3,6-TBA		H	2,3,6-trichlorobenzoic acid
TCA		H	trichloroacetic acid
TDE	DDD, Rhiothane	I	1,1-dichloro-2,2-di(4-chlorophenyl) ethane
Terbacil	Sinbar	H	3-*tert*-butyl-5-chloro-6-methyl-uracil
Terbutryn	Prebane, Igran	H	2-*tert*-butylamino-4-ethyl-amino-6-methylthio-*s*-triazine
TH-1568A	ACNQ		2-amino-3-chloro-1,4-naphthoquinone
Thiabend-azole	Mycozol	F	2-(thiazol-4-yl)benzimidazole
Thion-azin	Nemafos, Cynem	N	*O,O*-diethyl *O*-pyrazin-2-yl phos-phorothioate
Thioph-anate -methyl	Topsin-M	Fu	1,2-di-(3-methoxycarbonyl-2-thioureido)benzene
Thiram	Arasan Tersan	F	bis)dimethylthiocarbamoyl) disulfide

TABLE A-1 *(continued)*

Common name	Other name	Class	Chemical name
Toxaphene	Camphechlor	I	chlorinated camphene having a chlorine content of 67-69%
Triallate	Avadex BW	H	*S*-2,3,3-trichloroallyl diisopropylthiocarbamate
Tricamba	Banvel T	H	2,3,5-trichloro-6-methoxybenzoic acid
Trichl-oronat	Agritox, Agrisil	I,A	*O*-ethyl *O*-(2,4,5-trichlorophenyl)ethylphosphonothioate
Trietazine	G-27901	H	2-chloro-4-diethylamino-6-ethylamino-*s*-triazine
Trifluralin	Treflan	H	2,6-dinitro-*N*,*N*-dipropyl-4-trifluoromethylaniline
Vernolate	Vernam	H	*S*-propyl dipropylthiocarbamate
WL 41706	Fenproponate	I	α-cyano-3-phenoxybenzyl-2,2,3,3-tetramethyl cyclopropanecarboxylate
Zineb	Dithane-Z-78, Parzate	F	zinc-ethylene bisdithiocarbamate (of uncertain composition - polymeric)

[1] Most of the data given in this table were obtained from the Herbicide Handbook of the Weed Science Society of America, and Pesticide Mannual of British Crop Protection Council

Abbreviations - A = acaracid, F = fungicide, Fu - fumigant,
H = herbicide, I = insecticide, M = molluscicide,
N - nematicide

TABLE A-2

Some properties of pesticides referred to in text[1]

Pesticide	Physical state	M.P. (°C)	B.P. (°C)	Vapour pressure mm Hg (°C)	Solubility in water, ppm (°C)	LD_{50} mg/kg
Agvitor						100
Alachlor	S	39.5-41.5		0.02 (100)	148 (25)	1800
Aldicarb	S	100		0.05 (20)	6000	1
Aldrin	S	104-104.5		7.5×10^{-5} (20)	0.027 (25)	67
Ametryn	S	84-85		8.4×10^{-7} (20)	185 (20)	1110
Amiprophos						750
Amitrole	S	159			28×10^{4} (25)	26600
Arsenic trioxide						138
Atrazine	S	173-175		3.0×10^{-7} (20)	33 (27)	3080
Azinphosmethyl	S	73-74		3.8×10^{-4} (20)	33	16.4
Barban	S	75			11 (25)	1050
Benfluralin	S	65-66.5	121-122	4×10^{-7} (25)	< 1 (25)	800
Benomyl	S				Insol	10000
Bensulide	L-S	34.4			25 (20)	
Bentazon	S	137-139			500	1100
Bromacil	S	158-159		8×10^{-4} (100)	815 (25)	5200
Bromophos	S	53-54		1.3×10^{-4} (20)	40	3750
Bromoxynil	S	190			<200	250
Butralin	S	60-61	134-136		1.0	
Cacodylic acid	S	200			66.7×10^{4} (20)	830
Captan	S	178		$<1\times10^{-5}$ (25)	<0.5	9000
Carbaryl	S	142		<0.005 (26)	40 (30)	850
Carbendazim	P	307-312			8 (24)	1500
Carbofuran	S	150-152		2×10^{-5} (33)	700 (25)	8-14
Carbon disulphide	L	-108.6	46.3	357.1 (25)	2.2×10^{3} (32)	
Carbophenothion	L		82	3×10^{-7} (20)	<40	32
Carboxin	S	91.5-92.5			170 (25)	3820
CDAA	L		74	9.4×10^{-3} (20)		750
CDEC	L		128	2.2×10^{-3} (200)	92 (25)	850

Chloramben	S	201			700 (25)	3500
Chloranil	S	290			250	4000
Chlordane	L			$1x10^{-5}$ (25)	Insol	457-590
Chlorfenvinphos	L	-19	167-170	$4.0x10^{-6}$ (20)	145 (23)	10-39
Chloroneb	S	133-135		$3x10^{-3}$ (25)	8 (25)	>11000
Chlorphenamidine	S	32			250 (20)	127-352
Chloropicrin	L	-64	112.4	23.8 (25)	2270 (0)	
Chloroxuron	S	151-152			2.7	3700
Chlorpropham	S	38-40		$1x10^{-5}$ (25)	88	5000-7500
Chlorpyrifos	S	42.5-43		$1.87x10^{-5}$ (25)	2 (35)	163
Chlorthiamid	S	151-152		$1x10^{-6}$ (20)	950 (21)	757
Chlortoluron	S	147-148		$3.6x10^{-8}$ (20)	10 (20)	>10000
Crotoxyphos	L		135	$1.4x10^{-5}$ (20)		125
2,4-D	S	135-138	160	0.4 (160)	600 (20)	300-1000
Dalapon	L	185-190				
2,4-DB	S	120-121			46 (20)	700
DCPA	S	156		<0.01 (40)	0.5 (25)	>3000
2,4-DEP	L		200		Insol	850
DDT	S			$1.9x10^{-7}$ (20)	Insol	113
Demeton-*O*	L		123	$2.48x10^{-4}$	60 (22)	30
DEXON						60
Diallate	L		150		14	395
Diazinon	L		83-84	$1.4x10^{-4}$	40 (22)	300-850
Dibromochloropropane	L		196	0.8	$1x10^{-4}$	170-300
Dicamba	S	114-116		$3.7x10^{-3}$	$45x10^{2}$	2900
Dichlobenil	S		270	$5.5x10^{-4}$	18	3160
Dichlofenthion	L		120-123	0.2	0.3 (25)	270
Dichlormate	S	52		N (25)	170 (25)	1870-2140
Dichloropropene mixture	L		104		10^{3} (20)	250-500
Dichlorvos	L		35	$1.2x10^{-2}$	$1x10^{4}$	80
Dicrotophos			400	$1x10^{-4}$ (20)	M	16.5-22
Dicryl	S	121-126			Insol	3160
Dieldrin	S	175-176		$3.1x10^{-6}$ (20)	0.19 (25)	46
Dimefox	L		67	0.36 (25)	M	1-2
Dimethoate	S	51-52		$8.5x10^{-6}$ (25)	$2.5x10^{4}$	500-600
Dinitramine	S	98-99		$3.6x10^{-6}$ (25)	1.1 (25)	

TABLE A-2 *(continued)*

Pesticide	Physical state	M.P. (^{o}C)	B.P. (^{o}C)	Vapour pressure mm Hg (^{o}C)	Solubility in water, ppm (^{o}C)	LD_{50} mg/kg
Dinoseb	S	32		1 (151.1)	52 (25)	5-60
Diphenamid	S	132-135.5			261 (27)	686-776
Diquat	S				$70x10^{4}$	231
Disodium methane-arsenic acid						750
Disulfoton	L		62	$1.8x10^{4}$ (20)	25 (22)	8.6
Diuron	S	158-159		$3.1x10^{-6}$ (50)	42 (25)	3400
DMPA	S	51			5 (25)	270
DNOC	S	86		$1.05x10^{-4}$	130 (15)	25-40
DSMA	S	132-139			$25.4x10^{4}$	1800
Endosulfan	S	70-100		$1x10^{-5}$ (25)	Insol	80-110
Endothall	S	144			$10x10^{4}$ (20)	38-51
Endrin	S			$2x10^{-7}$ (25)	Insol	7.5-17.5
EPBP	L					274
EPTC	L		235	$34x10^{-3}$ (25)	370 (20)	1652
Ethion	L			$1.5x10^{-6}$ (25)	sol-sl	208
Ethirimol	S	159-160		$2x10^{-6}$ (25)	200 (22)	4000
Ethylene Dibromide	L		131.5	11.0 (25)	430 (30)	146
Fenac	S	157-160			sol-sl	1780
Fenitrothion	L		140-145	$6x10^{-6}$ (20)	Insol	250-500
Fensulfothion	L		138-141		154 (25)	4.6-10.5
Fenthion	L		87	$3x10^{-5}$ (20)	56 (22)	190-315
Fenuron	S		133-134	$1.6x10^{-4}$ (60)	$38.5x10^{2}$ (25)	6400
Fenvalerate	L					450
Fluchloralin	S	42-43				1550
Fluometuron	S	163-164.5			90 (25)	8900
Fonofos	L		130	$2.1x10^{-4}$	Insol	8-17.5
Glyphosate	S	230			$1.2x10^{4}$ (25)	4320
Heptachlor	S	95-96		$3x10^{-4}$ (25)	0.05 (25)	100
Hexachlorobenzene	S	226		$1.089x10^{-5}$ (20)	Insol	10000
Ioxynil	S	212-213.5			50 (25)	110
Leptophos	S	70.2-70.6			2.4 (25)	50

Lindane	S	159-160 (α-isomer)		0.06 (40)	Insol	
Linuron	S	93-94		1.5×10^{-5} (24)	75 (25)	4000
Malathion	L	2.85	156-157	4×10^{-5} (30)	145 (22)	2800
Maneb	S				sol-sl	6750
MCPA	S	118-119			M	700
Mecarbam	L	144			$<1 \times 10^{3}$	106
Metacrephos	L				10-30 (20)	630-790
Methabenzthiazuron	S	119-120		1×10^{-6} (20)	59 (20)	>2500
Metham	S				72.2×10^{4} (20)	285
Methidathion	S	39-40		1×10^{-6} (20)	240 (25)	25-54
Methiocarb	S	117-118			Insol	100
Methomyl	S	78-79		5×10^{-5} (25)	5.8×10^{4} (25)	17-24
Methyl Bromide	G		4.5		1.34×10^{4}	
Methoxyclor	P	89 (*p,p*-isomer)			Insol	600
Methylmercury dicyandiammide	S	156-157		6.5×10^{5} (35)	2.2×10^{4} (22)	
Metobromuron	S	95.5-96			330 (20)	2000
Mevinphos	L	99-103			M	3-12
Mocap	L	86-91		3.5×10^{-4} (26)	750	62
Molinate	L	202		5.6×10^{-3} (20)	800 (20)	720
Monolinuron	S	79-80		1.5×10^{-4} (22)	580	2250
Monuron	S	174-175		5×10^{-7} (25)	230 (25)	3600
MSMA	L				M	900
Naptalam	S	185			200 (25)	>8200
Neburon	S	102-103			4.8 (28)	11000
Nitralin	S	151-152		$<1.5 \times 10^{-6}$ (25)	0.6 (25)	>2000
Norea	S	171-172			150 (25)	1476-6830
Oryzalin	S	137-138			85 (25)	>10000
Oxadiazon	S	90		$<10^{-6}$ (20)	0.7 (20)	8000
Oxamyl	S	100-102		2.3×10^{-4} (25)	28×10^{4} (25)	5.4
Oxycarboxin	S	127.5-130			1000 (25)	2000
Oxydisulfoton	L			6.29×10^{-8} (20)	100 (22)	3.5
Panogen[R]						25
Paraquat	S				sol	150
Parathion	L	157-162		3.78×10^{-5} (20)	24 (25)	13

TABLE A-2 *(continued)*

Pesticide	Physical state	M.P. (^{o}C)	B.P. (^{o}C)	Vapour pressure mm Hg (^{o}C)	Solubility in water, ppm (^{o}C)	LD_{50} mg/kg
Parathion Methyl	S	35-36		0.97×10^{-5} (20)	60 (25)	14
PCP	S	191		0.12 (100)	30 (50)	27-80
PCNB	S	146				12000
Pebulate	L	142.5		6.8×10^{-2} (30)	60 (20)	1120
Phenthoate	S	17.5			11 (24)	300-400
Phenylmercury acetate	S	149-153		9×10^{-6} (35)	4370	
Phorate	L		118-120	8.4×10^{-4} (20)	50	3.7
Phosalone	S	48			10	120
Phosfolan	S	37-45			sol	8.9
Picloram	S			6.16×10^{-7} (35)	430 (25)	8200
Pirimicarb	S	90.5		3×10^{-5} (30)	27×10^{2} (25)	147
Profenofos	L		110			400
Profluralin	S	27-28			0.1 (25)	10000
Prometone	S	91-92		2.3×10^{-6} (20)	750 (20)	2980
Prometryn	S	118-120		1.0×10^{-6} (20)	48 (20)	3750
Pronamide	S	154-156		8.5×10^{-5} (25)	15 (15)	8350
Propachlor	S	67-76		0.03 (110)	700 (20)	710
Propanil	S	92-93		9×10^{-5} (60)	225	1300-1500
Propazine	S	212-214		2.9×10^{-8} (20)	8.6 (20)	>5000
Propham	S	87-88				
Pyrazon	S	207		0.074 (40)	300 (20)	3000
Pyrichlor						80
Quintozene	S	146		133×10^{-4} (25)	Insol	>12000
Ronnel	P	40-42		8×10^{-4} (25)	40 (22)	1740
Sesone	S	245			26.5×10^{4} (25)	1400
Siduron	S	133-138		$<8 \times 10^{-4}$ (100)	18 (25)	>5000
Simazine	S	225-227		6.1×10^{-9} (20)	5 (20)	>5000
Sodium Arsenite	S				sol	10-50
Solan	S	82-86			Insol	>10000
Swep	S	112-114			Insol	522

2,4,5-T	S	154-155	low	238 (30)	300
2,3,6-TBA	S	125-126		$8.4x10^{3}$ (20)	750-1000
TCA	S	59	5 (77)	$13x10^{6}$ (25)	5000
Terbacil	S	175-177	$4.8x10^{-7}$ (29.5)	710 (25)	>5000 <7500
Terbutryn	S	104	$9.6x10^{-7}$ (20)	58 (20)	2400-2980
Thiabendazole	P	304-305		<50 (25)	3330
Thionazin	L	-1.67	$3x10^{-3}$ (30)	1140 (27)	12
Thiophamatemethyl	S	172			>6000
Thiram	S	155-156		30(22)	375-865
Toxaphene			0.2-0.4 (25)	3 (22)	90
Triallate	L	148-149		4 (25)	1675-2165
Tricamba	S	137-139		sol-sl	970
Trichloronat	L	108		50 (20)	16-40
Trietazine	S	102-104		20 (25)	2830
Trifluralin	S	48.5-49	$1.99x10^{-4}$ (29.5)	<1 (27)	3700
Vernolate	L	150 (30)	$10.4x10^{-3}$ (25)	90 (20)	1780
Zineb	P			10 (22)	5200

[1] LD_{50} values refer to rats; however a few values apply to mice

Abbreviations - B.P. = boiling point, M.P. = melting point, G = gas, L = liquid, P = powder, S = solid, inscl = insoluble, M = miscible, sol = soluble, sol-m = soluble moderate, sol-sl = soluble slight.

AUTHOR INDEX

Underlined numbers give the page on which the complete reference is listed

SUBJECT INDEX